PURE AND APPLIED MATHEMATICS

A Program of Monographs, Textbooks, and Lecture Notes

Executive Editors — *Monographs, Textbooks, and Lecture Notes*

Earl J. Taft
Rutgers University
New Brunswick, New Jersey

Edwin Hewitt
University of Washington
Seattle, Washington

Chairman of the Editorial Board

S. Kobayashi
University of California, Berkeley
Berkeley, California

Editorial Board

Masanao Aoki
University of California, Los Angeles

Glen E. Bredon
Rutgers University

Sigurdur Helgason
Massachusetts Institute of Technology

G. Leitman
University of California, Berkeley

W. S. Massey
Yale University

Irving Reiner
University of Illinois at Urbana-Champaign

Paul J. Sally, Jr.
University of Chicago

Jane Cronin Scanlon
Rutgers University

Martin Schechter
Yeshiva University

Julius L. Shaneson
Rutgers University

LECTURE NOTES IN PURE AND APPLIED MATHEMATICS

1. *N. Jacobson*, Exceptional Lie Algebras
2. *L.-Å. Lindahl* and *F. Poulsen*, Thin Sets in Harmonic Analysis
3. *I. Satake*, Classification Theory of Semi-Simple Algebraic Groups
4. *F. Hirzebruch, W. D. Neumann,* and *S. S. Koh*, Differentiable Manifolds and Quadratic Forms
5. *I. Chavel*, Riemannian Symmetric Spaces of Rank One
6. *R. B. Burckel*, Characterization of C(X) among Its Subalgebras
7. *B. R. McDonald, A. R. Magid,* and *K. C. Smith*, Ring Theory: Proceedings of the Oklahoma Conference
8. *Yum-Tong Siu*, Techniques of Extension of Analytic Objects
9. *S. R. Caradus, W. E. Pfaffenberger,* and *Bertram Yood*, Calkin Algebras and Algebras of Operators on Banach Spaces
10. *Emilio O. Roxin, Pan-Tai Liu,* and *Robert L. Sternberg*, Differential Games and Control Theory
11. *Morris Orzech* and *Charles Small*, The Brauer Group of Commutative Rings
12. *S. Thomeier*, Topology and its Applications
13. *Jorge M. López* and *Kenneth A. Ross*, Sidon Sets
14. *W. W. Comfort* and *S. Negrepontis*, Continuous Pseudometrics
15. *Kelly McKennon* and *Jack M. Robertson*, Locally Convex Spaces
16. *M. Carmeli* and *S. Malin*, Representations of the Rotation and Lorentz Groups: An Introduction
17. *George B. Seligman*, Rational Methods in Lie Algebras
18. *Djairo Guedes de Figueiredo*, Functional Analysis: Proceedings of the Brazilian Mathematical Society Symposium

Other volumes in preparation

Functional Analysis

PROCEEDINGS OF THE BRAZILIAN
MATHEMATICAL SOCIETY SYMPOSIUM
CAMPINAS, BRAZIL

edited by

Djairo Guedes de Figueiredo

Departamento de Matemática
Instituto de Ciencias Exatas
Universidade de Brasília
Brasília, Brazil

MARCEL DEKKER, INC.   New York and Basel

COPYRIGHT © 1976 by MARCEL DEKKER, INC. ALL RIGHTS RESERVED.

Neither this book nor any part may be reproduced or transmitted in any form or by any means, electronic or mechanical, including photocopying, microfilming, and recording, or by any information storage and retrieval system, without permission in writing from the publisher.

MARCEL DEKKER, INC.

270 Madison Avenue, New York, New York 10016

LIBRARY OF CONGRESS CATALOG CARD NUMBER: 75-12363

ISBN: 0-8247-6334-3

Current printing (last digit):
10 9 8 7 6 5 4 3 2 1

PRINTED IN THE UNITED STATES OF AMERICA

CONTENTS

Preface vii

ITERATES OF ELLIPTIC OPERATORS AND BERNSTEIN INEQUALITIES   1
   M. S. Baouendi

HEWITT'S STONE-WEIERSTRASS THEOREMS FOR ORDERED TOPOLOGICAL
SPACES   9
   Jörg Blatter

CLASSES OF DISTRIBUTIONS WITH PARABOLIC MAXIMAL FUNCTIONS
IN $L^p(R^n)$ 27
   Alberto P. Calderón and Alberto Torchinsky

LOCALLY SOLVABLE PSEUDODIFFERENTIAL OPERATORS WITH DOUBLE
CHARACTERISTICS   39
   Fernando Cardoso

ENERGY CONCENTRATION FOR FIRST-ORDER HYPERBOLIC SYSTEMS   53
   David G. Costa

EXISTENCE THEOREMS IN HIGHER DIMENSIONAL CALCULUS OF
VARIATIONS   71
   Ubiratan D'Ambrosio

ON THE LOCAL SOLVABILITY OF LINEAR PARTIAL DIFFERENTIAL
OPERATORS   79
   Antonio Gilioli

ON STRONGLY NONLINEAR ELLIPTIC PROBLEMS   91
   Peter Hess

UNFOLDING OF SINGULARITIES   111
   Peter Hilton

THE DIRICHLET AND SUBSTITUTION FORMULAS FOR RIEMANN-
STIELTJES INTEGRALS IN BANACH SPACES   135
   Chaim Samuel Hönig

FUNCTIONAL DIFFERENTIAL EQUATIONS 191
   A. F. Izé and O. F. Lopes

ON THE GEOMETRY AND STRUCTURE OF BANACH SPACES 215
   L. A. Karlovitz

ANALYTIC TRANSFORMATION GROUPS 235
   Wilhelm Kaup

ON LOCALLY CONVEX SPACES WITH THE MONTEL PROPERTY 245
   Mário C. Matos

SOME HOLOMORPHICALLY SIGNIFICANT PROPERTIES OF LOCALLY
CONVEX SPACES 251
   Leopoldo Nachbin

REMARKS ON NONLINEAR EVOLUTION EQUATIONS IN NONCYLINDRICAL
DOMAINS 279
   B. P. Neves and L. A. Medeiros

SOLUTIONS OF A NONLINEAR ABSTRACT CAUCHY-KOVALEWSKY
SYSTEM AS A LOCAL BANACH ANALYTIC MANIFOLD 291
   Domingos Pisanelli

WEIGHTED APPROXIMATION AND FULL COMPLETENESS IN SPACES
OF CONTINUOUS FUNCTIONS 301
   W. H. Summers

AN INVITATION TO THE THEORY OF ORDINARY DIFFERENTIAL
EQUATIONS IN BANACH SPACES 309
   Giovanni Vidossich

Index 323

LIST OF CONTRIBUTORS

M. S. BAOUENDI, Department of Mathematics, Purdue University, West Lafayette, Indiana

JÖRG BLATTER,[*] Institut für Angewandte Mathematik, Universität Bonn, Bonn, Federal Republic of Germany and Instituto de Matemática, Universidade Federal do Rio de Janeiro, Rio de Janeiro, Brazil

ALBERTO P. CALDERÓN,[†] Department of Mathematics, Massachusetts Institute of Technology, Cambridge, Massachusetts

FERNANDO CARDOSO, Instituto de Matemática, Universidade Federal de Pernambuco, Recife, Pernambuco, Brazil and The Institute for Advanced Study, Princeton, New Jersey

DAVID G. COSTA,[‡] Instituto de Matemática, Universidade Federal do Rio de Janeiro, Rio de Janeiro, Brazil

UBIRATAN D'AMBROSIO, Instituto de Matemática, Universidade Estadual de Campinas, Campinas, São Paulo, Brazil

ANTONIO GILIOLI, Instituto de Matemática e Estatística, Universidade de São Paulo, São Paulo, Brazil

PETER HESS, Mathematisches Institut, Universität Zürich, Zürich, Switzerland

PETER HILTON, Battelle Research Center, Seattle, Washington and Department of Mathematics, Case Western Reserve University, Cleveland, Ohio

---

[*] Present address: Instituto de Matemática Pura e Aplicada, Rio de Janeiro, Brazil

[†] Present address: Department of Mathematics, University of Chicago, Chicago, Illinois

[‡] Present address: Departamento de Matemática, Universidade de Brasília, Brasília, Brazil.

CHAIM SAMUEL HONIG, Instituto de Matemática e Estatística, Universidade de São Paulo, São Paulo, Brazil

A. F. IZÉ, Instituto de Ciências Matemática de São Carlos, Departamento de Matemática, Universidade de São Paulo, São Carlos, São Paulo, Brazil

L. A. KARLOVITZ, Institute for Fluid Dynamics and Applied Mathematics, University of Maryland, College Park, Maryland

WILHELM KAUP,[*] Departamento de Matemática, Universidade Federal do Rio de Janeiro, Rio de Janeiro, Brazil

O. F. LOPES, Instituto de Matemática, Estatística e Ciências de Computação, Universidade Estadual de Campinas, Campinas, São Paulo, Brazil.

MÁRIO C. MATOS, Departamento de Matemática do IMECC, Universidade Estadual de Campinas, Campinas, São Paulo, Brazil

L. A. MEDEIROS, Instituto de Matemática, Universidade Federal do Rio de Janeiro, Rio de Janeiro, Brazil

LEOPOLDO NACHBIN, Instituto de Matemática, Universidade Federal do Rio de Janeiro, Rio de Janeiro, Brazil

B. P. NEVES, Instituto de Matemática, Universidade Federal do Rio de Janeiro, Rio de Janeiro, Brazil

DOMINGOS PISANELLI, Departamento de Matemática, Universidade de São Paulo, São Paulo, Brazil

W. H. SUMMERS,[†] Instituto de Matemática, Universidade Federal do Rio de Janeiro, Rio de Janeiro, Brazil

ALBERTO TORCHINSKY,[‡] Department of Mathematics, Cornell University, Ithaca, New York

GIOVANNI VIDOSSICH,[#] Departamento de Matemática, Universidade de Brasília, Brasília, Brazil

---

[*]Present address: Mathematisches Institut de Universität Tübingen, Tübingen, Germany

[†]Present address: Department of Mathematics, University of Arkansas, Fayetteville, Arkansas

[‡]Present address: Department of Mathematics, Indiana University, Bloomington, Indiana

[#]Present address: Instituto di Matematica, Università di Trieste, Trieste, Italy.

PREFACE

This book contains the texts of some of the lectures delivered at the Symposium of Functional Analysis held at Instituto de Matemática da Universidade Estadual de Campinas, São Paulo, Brazil, in July, 1974.

This meeting was supported by Conselho Nacional do Desenvolvimento Científico e Tecnológico (CNPq), Coordenação de Aperfeiçoamento do Pessoal de Ensino Superior (CAPES), Fundação do Amparo a Pesquisa do Estado de São Paulo (FAPESP), Organization of the American States (OAS), and was sponsored by the Sociedade Brasileira de Matemática.

There were ten courses, which were delivered by Alberto Calderón (H^p-parabolic spaces), Antonio Fernando Izé and Orlando Lopes (Functional differential equations), Carlos Isnard (Introduction to degree theory), Chaim Samuel Hönig (Functional analysis and integro-differential equations with linear constraints), Djairo Guedes de Figueiredo (Introduction to monotone operator theory), Geraldo Avila (Topics in wave propagation), Leopoldo Nachbin (Infinite dimension holomorphy), Les Karlovitz (Topics in the structure and geometry of Banach spaces), Pedro Nowosad (Extremal problems for eigenvalue), and Peter Hess (On strongly nonlinear elliptic problems), seventeen invited lectures, and a number of research announcements. The number of participants exceeded 90.

The Organizational Committee was constituted by Professors Chaim Samuel Hönig, Djairo Guedes de Figueiredo and Pedro Nowosad.

We would like to thank the University of Campinas administration for the excellent logistic support given to us during the realization of the meeting. Finally a word of acknowledgment to Mr. Newton de Góes Horta for the expert typing of the manuscripts that compose the present volume and the material of the courses distributed at the time of the conference.

Brasília
July, 1976   Djairo Guedes de Figueiredo

ITERATES OF ELLIPTIC OPERATORS AND BERNSTEIN INEQUALITIES

M. S. Baouendi[†]

Department of Mathematics
Purdue University
West Lafayette, Indiana

We want to show in this chapter that the classical Bernstein inequality, estimating any polynomial P in the complex plane by means of the supremum of $|P|$ on $[-1, 1]$, can be extended to the eigenfunctions of a suitable elliptic operator A. Such inequalities imply also some asymptotic behavior of the spectrum of A, and are equivalent to the analyticity of functions satisfying inequalities involving the iterates of A.

1. NOTATIONS AND ASSUMPTIONS

Let Ω be an open bounded set of $\mathbb{R}^n$ with Lipschitzian boundary, and A a differential operator of order 2 with

---

[†]Supported by NSF Grant GP 35825. Work done jointly with C. Goulaonic, University of Paris XI.

coefficients in $A(\overline{\Omega})$, space of analytic functions on $\overline{\Omega}$ (closure of Ω). We assume that A is formally self-adjoint. We denote by $(A, \mathcal{D}(A))$ a strictly positive self-adjoint realization of A in $L^2(\Omega)$. We assume that the embedding of $\mathcal{D}(A)$ into $L^2(\Omega)$ is compact. We denote by (λ_j), $j \in \mathbb{N}$, the nondecreasing sequence of eigenvalues of A repeated according to their multiplicity, and by (φ_j) the sequence of eigenfunctions associate to (λ_j). We assume that the φ_j's are in $A(\overline{\Omega})$.

We denote by $\mathcal{D}(A^k)$ the domain of A^k defined recursively by $\mathcal{D}(A^k) = \{u \in \mathcal{D}(A^{k-1}), Au \in \mathcal{D}(A^{k-1})\}$. We denote

$$\mathcal{D}(A^\infty) = \bigcap_{k \in \mathbb{N}} \mathcal{D}(A^k).$$

We suppose that for large enough $r \in \mathbb{N}$, $\mathcal{D}(A^r)$ consists of continuous functions on $\overline{\Omega}$.

Finally, we assume that $\mathcal{D}(A^\infty) \cap A(\overline{\Omega})$ is a closed subspace of $A(\overline{\Omega})$ equipped with its usual inductive limit topology.

All the assumptions stated here are satisfied in many cases (see the examples given later). We suppose them in the rest of this discussion.

2. THE "ITERATE PROPERTY" AND THE ASYMPTOTIC BEHAVIOR OF THE EIGENVALUES

We denote

$$\mathcal{D}(A^\omega) = \{u \in \mathcal{D}(A^\infty); \exists L > 0, \forall k \in \mathbb{N}, \|A^k u\|_{L^2(\Omega)} \leq L^{k+1}(2k)!\}$$

equipped with its natural inductive limit topology. In fact, one changes neither the space nor its topology by replacing the L^2 norm by the supremum norm.

It is not difficult to see that we always have

$$\mathcal{D}(A^\infty) \cap A(\overline{\Omega}) \subset \mathcal{D}(A^\omega).$$

DEFINITION. The operator A satisfies the "iterate property" iff

$$\mathcal{D}(A^\omega) = \mathcal{D}(A^\infty) \cap A(\overline{\Omega}). \tag{1}$$

THEOREM 1. If A satisfies the "iterate property" (1), there are two positive constants C_1 and C_2 such that, for every $j \in \mathbb{N}$,

$$C_2 j^{2/n} \leq \lambda_j \leq C_1 j^{2/n}. \tag{2}$$

The inequality

$$\lambda_j \leq C_1 j^{2/n}$$

is easy and follows from the fact that the order of A is 2. The proof of the inequality

$$\lambda_j \geq C_2 j^{2/n}$$

is more complicated. We do not give it here (see [1] for more details); we only mention that it is based on the following lemma, which we also state without proof.

If $\alpha = (\alpha_j)$ is a sequence of nonnegative real numbers and $b > 0$, we denote

$$\ell^2(b, \alpha) = \{(c_k) \in \mathbb{C}^\mathbb{N}; \sum_{k=0}^{\infty} |c_k|^2 b^{\alpha_k} < \infty\}$$

and

$$E(\alpha) = \underset{b>1}{\text{ind lim}} \; \ell^2(b, \alpha).$$

LEMMA. Let $\alpha = (\alpha_j)$ and $\beta = (\beta_j)$ be two nondecreasing sequences of strictly positive real numbers satisfying $\lim_{j\to\infty} \alpha_j = \lim \beta_j = \infty$. If $E(\alpha)$ is isomorphic to a closed subspace of $E(\beta)$, then there is $C > 0$ such that

$$\beta_j \leq C\alpha_j \quad \text{for every } j \in \mathbb{N}.$$

3. BERNSTEIN INEQUALITY

We state first the main theorem:

THEOREM 2. The following properties are equivalent:

(i) The operator A satisfies the "iterate property" (1).

(ii) For every real number $b > 1$ there is a neighborhood V of $\overline{\Omega}$ in $\mathbb{C}^n$ such that, for every $j \in \mathbb{N}$, φ_j can be extended analytically in V and the following inequality holds:

$$\sup_{x \in V} |\varphi_j(x)| \leq b^{\sqrt{\lambda_j}} \sup_{x \in \Omega} |\varphi_j(x)|.$$

(iii) For every real number $b > 1$ there is a neighborhood V of $\overline{\Omega}$ in $\mathbb{C}^n$ such that, for every $j \in \mathbb{N}$, every $f \in H_j$ (finite dimensional space spanned by $\varphi_0, \ldots, \varphi_j$) can be extended analytically in V and the following inequality holds:

$$\sup_{x \in V} |f(x)| \leq b^{\sqrt{\lambda_j}} \sup_{x \in \Omega} |f(x)|.$$

The proof of this theorem is essentially based on the following more abstract one proved in [2]:

Let (H_j) be a sequence of finite dimensional vector spaces satisfying

$$H_0 \subset H_1 \subset \cdots \subset H_j \subset H_{j+1} \subset \cdots \subset A(\overline{\Omega}).$$

Let μ_j be an increasing sequence of strictly positive real numbers such that, for every $a \in (0, 1)$,

$$\sum_{j=0}^{\infty} a^{\mu_j} < \infty.$$

We introduce the following properties:

<u>Property (A)</u>. Any function $f \in C(\overline{\Omega})$ for which there exist $C > 0$ and $a \in (0, 1)$ such that for each $j \in \mathbb{N}$

$$d(f, H_j) = \inf_{g \in H_j} \sup_{x \in \Omega} |f(x) - g(x)| \leq Ca^{\mu_j}$$

belongs to $A(\overline{\Omega})$.

<u>Property</u> $[B(\mu_j)]$. For any real number $b > 1$ there exists an open set V, neighborhood of $\overline{\Omega}$ in $\mathbb{C}^n$ such that any $g \in H_j$ can be extended analytically to V and

$$\sup_{x \in V} |g(x)| \leq b^{\mu_j} \sup_{x \in \Omega} |g(x)|.$$

THEOREM 3. Property (A) implies property $[B(\mu_j)]$. Property $[B(\mu_{j-1})]$ implies property (A).

The proof of Theorem 3 is based on the Baire property. With $\mu_j = \sqrt{\lambda_j}$, where (λ_j) satisfies Eq. (2), it is easily seen that $[B(\mu_j)]$ is equivalent to $[B(\mu_{j-1})]$.

4. EXAMPLES

1. We take here $n = 1$, $\Omega = (-1, 1)$, $A = -(d/dx)(1 - x^2)(d/dx) + 1$ (Legendre operator). The eigenfunctions are the Legendre polynomials and $\lambda_j = j(j + 1) + 1$.

$$D(A) = \{u \in L^2((-1, +1)), u' \in L^2((-1, +1)),$$
$$(1 - x^2)u'' \in L^2((-1, +1))\}.$$

In this case, A satisfies statement (iii) of Theorem 2 (classical Bernstein inequality; see [3, 4]) and hence A satisfies the "iterate property." It should be noted that we have in this case

$$D(A^\omega) = A([-1, 1]).$$

2. Let $\overline{\Omega}$ be an analytic manifold with boundary. $A = -\Delta$ (Laplace operator), $D(A) = H^2(\Omega) \cap H_0^1(\Omega)$, $H^m(\Omega)$ and $H_0^m(\Omega)$ being the usual Sobolev spaces (see [5] for example). A satisfies the "iterate properties" (see [5, 6]) and hence satisfies statements (ii) and (iii) of Theorem 2.

More general elliptic boundary problems may be considered.

3. Let φ be an analytic real function defined in a neighborhood of $\overline{\Omega}$ such that

$$\Omega = \{x;\ \varphi(x) > 0\}$$

$$d\varphi(x) \neq 0 \quad \text{for} \quad x \in \partial\Omega.$$

$$A = -\text{div } \varphi \text{ grad } \varphi + I,$$

$$D(A) = \{u \in H^1(\Omega), \varphi u \in H^2(\Omega)\}.$$

It is proved in [7] that A does not satisfy Eq. (2) and therefore A does not satisfy the "iterate property."

4. We use the same notations as in the previous example. We denote

$$\Lambda_{ij} = \frac{\partial \varphi}{\partial x_i} \frac{\partial}{\partial x_j} - \frac{\partial \varphi}{\partial x_j} \frac{\partial}{\partial x_i},$$

$$A = -\text{div } \varphi \text{ grad } \varphi + \sum_{i,j} \Lambda_{i,j}^* \Lambda_{i,j} + 1$$

($\Lambda_{i,j}^*$ being the adjoint of $\Lambda_{i,j}$).

$$\mathcal{D}(A) = \{u \in H^1(\Omega), \varphi u \in H^2(\Omega), \Lambda_{i,j}^2 u \in L^2(\Omega)\}.$$

It is proved in [8] that A satisfies the "iterate property" and hence the eigenfunctions satisfy the Bernstein inequality, that is, conditions (ii) and (iii) of Theorem 2. As in Example 1, it should be observed that we have here

$$\mathcal{D}(A^\omega) = A(\overline{\Omega}).$$

<u>Remark</u>. All the results stated here can be extended without any difficulty to the case of operators of order higher than 2. We have chosen the second-order case only for the sake of simplicity.

REFERENCES

1. M. S. Baouendi and C. Goulaouic, Itérés d'opérateurs elliptiques et prolongement de fonctions propes, *Rev. Roumaine Math.*, 18 (1973), 1495-1501.
2. M. S. Baouendi and C. Goulaouic, Approximation of analytic functions on compact sets and Bernstein's inequality, *Trans. Amer. Math. Soc.*, 189 (1974), 251-261.
3. S. Bernstein, Collected works, Izdat. Akad. Nauk. SSSR, Moscow, 1952.

4. S. N. Mergelian, Uniform approximations to functions of a complex variable, *Usp. Mat. Nauk.*, 7 (1952), no. 2 (48), 31-122, English transl. A. M. S. Transl. (1)3(1962), 294-391.

5. J. L. Lions and E. Magenes, *Problèmes aux limites non homogènes*, Dunod, Paris, 1970.

6. N. Aronszajn, Sur les décompositions des fonctions analytiques uniformes et sur leurs applications, *Acta Math.*, 65 (1935), 1-156.

7. M. S. Baouendi and C. Goulaouic, Régularité et théorie spectrale pour une classe d'opérateurs elliptiques dégénérés, *Arch. Rat. Mec. Anal.*, 34 (1969), 361-379.

8. M. S. Baouendi, C. Goulaouic, and B. Hanouzet, Caractérisation de classes de fonctions C^∞ et analytiques, *J. Math. pures et appl.*, 52 (1973), 115-144.

HEWITT'S STONE-WEIERSTRASS THEOREMS FOR ORDERED TOPOLOGICAL SPACES *

Jörg Blatter[†]

Institut für Angewandte Mathematik
Universität Bonn
Bonn, Federal Republic of Germany

and

Instituto de Matemática
Universidade Federal do Rio de Janeiro
Rio de Janeiro, Brazil

Let X be a completely regular Hausdorff topological space. Then for any uniformly closed algebra A of bounded continuous real-valued functions on X which contains the constant functions, each of the following conditions implies the next (the separation conditions are defined below).

A is the algebra of all bounded continuous real-valued functions on X.

A separates the zero sets of X.

---

*This work was supported by the Conselho Nacional de Pesquisas, Rio de Janeiro, and the Gesellschaft für Mathematik und Datenverarbeitung, Bonn.

[†]Present address: Instituto de Matemática Pura e Aplicada, Rio de Janeiro, Brazil.

A separates points and closed subsets of X.

A separates the points of X.

The classical Stone-Weierstrass theorem [1, 2] states that if X is compact, it has the property

> There exists only one uniformly closed algebra of bounded continuous real-valued functions on X which contains the constant functions and separates the points of X.

Hewitt [3] showed that this property actually characterizes compactness of X, found several conditions on X (see below) equivalent to the intermediate property

> There exists only one uniformly closed algebra of bounded continuous real-valued functions on X which contains the constant functions and separates points and closed subsets of X,

and proved that X always has the property

> There exists only one uniformly closed algebra of bounded continuous real-valued functions on X which contains the constant functions and separates the zero sets of X.

The purpose of the present note is an extension of Hewitt's results to Nachbin's [4] completely regular ordered topological spaces.

1. ORDERED TOPOLOGICAL SPACES

In this section we record those definitions and results concerning ordered topological spaces which will be needed in the sequel. For more detailed information refer to [4-9].

1.1. DEFINITION. An *ordered topological space* is a triple $(X, \tau, \leq)$ consisting of a set X, a topology τ for X, and an *order* $\leq$ for X, i.e., a transitive, reflexive, antisymmetric binary relation $\leq$ for X, which is closed, i.e., $\leq$ is a closed subset of $X \times X$ when the latter is given the product topology.

We note that the topology of an ordered topological space is always Hausdorff, and we comment briefly on two extreme cases. If X is a set and τ a topology for X and if the order $\leq$ for X is taken to be equality, then $(X, \tau, \leq)$ is an ordered topological space iff τ is Hausdorff. Thus Hausdorff topological spaces are just those ordered topological spaces whose order is equality. If X is a set, τ a topology for X and $\leq$ a *total order* for X, i.e., an order $\leq$ for X such that $x \leq y$ or $y \leq x$ for any $x, y \in X$, then $(X, \tau, \leq)$ is an ordered topological space iff τ contains the $\leq$ *order topology* for X, i.e., the smallest topology for X which contains all sets of the form $\{y \in X: y < x \ (y \leq x \text{ and } y \neq x)\}$ and $\{y \in X: x < y\}$ for some $x \in X$.

Now let X be an ordered topological space (whenever possible we shall omit explicit mention of the topology of an ordered topological space and shall denote the order by $\leq$). We shall denote by $C^*(X)$ the set of all bounded continuous real-valued functions on X equipped with the usual pointwise algebra and lattice operations and with the topology of uniform convergence, and by $C_i^*(X)$ the set of *increasing* members of $C^*(X)$, i.e., members f of $C^*(X)$ such that $f(x) \leq f(y)$ whenever $x, y \in X$ and $x \leq y$. The sets of the form $\{x \in X: f(x) \geq 0\}$

(resp. $\{x \in X: f(x) \leq 0\}$) for some $f \in C_i^*(X)$ will be called the *nonnegative* (resp. *nonpositive*) *sets* of X. The nonnegative and nonpositive sets of X together are called the *zero sets* of X. A subset S of X is called *increasing* if $x \in S$ and $x \leq y$ imply $y \in S$ and *decreasing* if its complement is increasing. The set of increasing subsets of X is closed under the formation of arbitrary unions and intersections and so every subset S of X has a smallest increasing superset $i(S)$, its *increasing hull*, and a smallest decreasing superset $d(S)$, its *decreasing hull*; $i(X) = \{y \in X: x \leq y \text{ for some } x \in S\}$ and $d(S) = \{y \in X: y \leq x \text{ for some } x \in S\}$. A subset S of X is called *convex* if $x, y \in S$ and $x \leq z \leq y$ imply $z \in S$; equivalently, if it is the intersection of an increasing subset of X and a decreasing subset of X. The *convex hull* of a subset S of X is the smallest convex superset $i(S) \cap d(S)$ of S. The increasing hull and the decreasing hull of a compact subset of X are closed and thus also its convex hull.

1.2. DEFINITION. Let X be an ordered topological space and let A be a closed subalgebra of $C^*(X)$ which contains the constant functions. We shall denote by A_i the set of increasing members of A and shall say that

  (i) A_i *separates the points* of X if whenever x and y are points of X for which $x \leq y$ is false, there exists $f \in A_i$ such that $1_{\{x\}} \leq f \leq 1_{X \sim \{y\}}$ (= characteristic function of $X \sim \{y\}$),

  (ii) A_i *separates points and closed subsets* of X if whenever x is a point of X and F is a closed subset of X not containing x, there exist $f \in A_i$ and $g \in -A_i$ such that $1_{\{x\}} \leq f \wedge g \leq 1_{X \sim F}$,

  (iii) A_i *separates the zero sets* of X if whenever Z_1 is a nonnegative subset of X and Z_2 is a nonpositive subset of X disjoint from Z_1, there exists $f \in A_i$ such that $1_{Z_1} \leq f \leq 1_{X \sim Z_2}$.

Let us comment briefly on this definition. It is obvious that A_i separates the points of X iff it *determines the order* of X, i.e., for any two points x and y of X, $x \leq y$ iff $f(x) \leq f(y)$ for all $f \in A_i$, and it is only slightly less obvious that A_i separates points and closed subsets of X iff it *determines the topology* of X, i.e., the topology of X is the smallest topology for X which renders continuous the members of A_i. Simple examples show that A_i may separate points and closed subsets of X and not separate the points of X; it is not hard to see, however, that this cannot occur if the order of X is either equality or total. Finally, we note that $C_i^*(X)$ always separates the zero sets of X: Let $f, g \in C_i^*(X)$, set $Z_1 = \{x \in X: f(x) \geq 0\}$, $Z_2 = \{x \in X; g(x) \leq 0\}$, and assume that $Z_1 \cap Z_2 = \phi$. Then $(g \vee 0) - (f \wedge 0)$ is positive everywhere on X and $h = (g \vee 0)[(g \vee 0) - (f \wedge 0)]^{-1}$ is an element of $C_i^*(X)$ which satisfies $0 \leq h \leq 1$, $h^{-1}[\{1\}] = Z_1$ and $h^{-1}[\{0\}] = Z_2$.

1.3. DEFINITION. An ordered topological space X is called *completely regular* if $C_i^*(X)$ determines the topology and the order of X.

It is an easy exercise to verify that this definition can be paraphrased thus: An ordered topological space X is completely regular iff for any closed subalgebra A of $C^*(X)$ which contains the constant functions, each of the following conditions implies the next.

$A_i = C_i^*(X)$.

A_i separates the zero sets of X.

A_i separates points and closed subsets of X as well as the points of X.

A_i separates the points of X.

The completely regular ordered topological spaces whose order is equality are just the completely regular Hausdorff topological spaces. Call an ordered topological space *locally convex* if the neighborhood filter of any of its points has a base consisting of convex sets. Then: An ordered topological space is locally convex if — and if its order is total, also only if — it is completely regular.

1.4. DEFINITION. Let X be an ordered topological space. A *Nachbin compactification* of X is a pair (Y, κ) consisting of a *compact* ordered topological space Y, i.e., an ordered topological space Y whose topology is compact, and a map $\kappa: X \to Y$ such that

(1) κ is a topological embedding.

(2) κ is an *order embedding*, i.e.,

$$\kappa(x) \leq \kappa(y) \quad \text{iff} \quad x \leq y.$$

(3) $\kappa[X]$ is dense in Y.

(4) The order of Y is the smallest closed order for Y with respect to which κ is an order embedding.

We define a *preorder*, i.e., a transitive and reflexive binary relation, $\leq$ for the collection of all Nachbin compactifications of X by agreeing that $(Y_1, \kappa_1) \leq (Y_2, \kappa_2)$ iff there exists a map $\varphi: Y_2 \to Y_1$ which is continuous and satisfies $\varphi \circ \kappa_2 = \kappa_1$ (it follows that φ is increasing) and we call two Nachbin compactifications (Y_1, κ_1) and (Y_2, κ_2) *equivalent* if there exists a map $\varphi: Y_2 \to Y_1$ which is a homeomorphism and satisfies $\varphi \circ \kappa_2 = \kappa_1$ (it follows that φ is an order embedding); it follows that (Y_1, κ_1) and (Y_2, κ_2) are equivalent iff $(Y_1, \kappa_1) \leq (Y_2, \kappa_2)$ and $(Y_2, \kappa_2) \leq (Y_1, \kappa_1)$.

An ordered topological space has a Nachbin compactification iff it is completely regular, and a completely regular ordered topological space X (whenever possible considered as a subspace of any of its Nachbin compactifications so that, in particular, explicit mention of any embedding becomes unnecessary, and the equivalent ones among its Nachbin compactifications considered equal) has a unique Nachbin compactification $\beta_i X$ with the following equivalent properties:

(a) $Y \leq \beta_i X$ for any Nachbin compactification Y of X.

(b) For every $f \in C_i^*(X)$ there exists a unique $f\hat{\ } \in C_i^*(\beta_i X)$ which extends f, i.e., $f\hat{\ }|X = f$.

As a consequence of (b), for every $f \in c\ell[C_i^*(X) - C_i^*(X)]$ (= smallest closed subalgebra of $C^*(X)$ which contains $C_i^*(X)$!) there exists a unique $f\hat{\ } \in C^*(\beta_i X)$ which extends f and by the classical Stone-Weierstrass theorem

$$\{c\ell[C_i^*(X) - C_i^*(X)]\}\hat{\ } = C^*(\beta_i X).$$

If X is a completely regular ordered topological space whose order is equality, then the order of any Nachbin compactification of X is also equality, that is, the Nachbin compactifications of X are just its topological compactifications; in particular, $\beta_i X$ is the Stone-Čech compactification βX of X. If X is a completely regular ordered topological space whose order is total, then the order of any Nachbin compactification of X is also total and its topology is the corresponding order topology.

2. THE STONE-WEIERSTRASS THEOREMS

We commence with a theorem which, by now, is almost obvious. For the case that the order is equality see Hewitt [3; Theorems 1-2].

2.1. THEOREM. Let X be a completely regular ordered topological space and let A be a closed subalgebra of $C^*(X)$ which contains the constant functions. The following conditions are equivalent.

(a) $A_i = C_i^*(X)$

(b) A_i separates the zero sets of X.

(c) $(A_i)\hat{\ }$ separates the points of $\beta_i X$.

Proof. (a) implies (b). See the remark after 1.2.

(b) implies (a). Let x and y be points of $\beta_i X$ for which $x \leq y$ is false. Since $\beta_i X$ is a completely regular ordered topological space, there exists $f \in C_i^*(\beta_i X)$ such that $1_{\{x\}} \leq f \leq 1_{\beta_i X \sim \{y\}}$. $Z_1 = \{z \in X: f(z) \geq 2/3\}$ is a nonnegative set of X, $Z_2 = \{z \in X: f(z) \leq 1/3\}$ is a nonpositive set of X disjoint from Z_1, and $x \in c\ell(Z_1)$, $y \in c\ell(Z_2)$. Since A_i separates the zero sets of X, there exists $g \in A_i$ such that $1_{Z_1} \leq g \leq 1_{X \sim Z_2}$.

$$1_{\{x\}} \leq 1_{c\ell(Z_1)} \leq g\hat{\ } \leq 1_{\beta_i X \sim c\ell(Z_2)} \leq 1_{\beta_i X \sim \{y\}},$$

and so we are done with this implication.

(c) implies (a). By the classical Stone-Weierstrass theorem $c\ell[(A_i)\hat{\ } - (A_i)\hat{\ }] = C^*(\beta_i X)$ and thus $c\ell(A_i - A_i) \cap C_i^*(X) = C_i^*(X)$. Since $c\ell(A_i - A_i) \subset A$, $c\ell(A_i - A_i) \cap C_i^*(X) = A_i$ and this completes the proof.

Before we continue, we feel obliged to say a word or two about a consequence of the classical Stone-Weierstrass theorem, which we have actually used twice already. If X is a compact ordered topological space and if A is a closed subalgebra of $C^*(X)$ which contains the constant functions, then $A_i = C_i^*(X)$ iff $A = C^*(X)$, that is, the order of X plays, as far as our problem is concerned, no role whatsoever. We therefore recall the fact that for a noncompact completely regular ordered topological space X, there may very well exist a proper closed subalgebra A of $C^*(X)$ which contains the constant functions and satisfies $A_i = C_i^*(X)$, namely iff $c\ell[C_i^*(X) - C_i^*(X)] \neq C^*(X)$, and this is the case iff $\beta_i X$ and βX are topologically distinct.

2.2. THEOREM. Let X be a completely regular ordered topological space. The following conditions are equivalent.

(a) $A_i = C_i^*(X)$ whenever A is a closed subalgebra of $C^*(X)$ which contains the constant functions such that A_i separates points and closed subsets of X as well as the points of X.

(b) X has a unique Nachbin compactification.

(c) A convex subset S of $\beta_i X$ which is contained in $\beta_i X \sim X$ and satisfies

$x \leq y$ whenever $x, y \in X$ and $x \in d(S)$, $y \in i(S)$,

has at most one element.

Proof. (a) implies (b). Let Y be any Nachbin compactification of X and set $C = \{f \in C^*(Y): f|X \in C_i^*(X)\}$. $A = \{f|X: f \in C^*(Y)\}$ is a closed subalgebra of $C^*(X)$ which contains the constant functions, $A_i = \{f|X: f \in C\}$, and the latter set separates points and closed subsets of X as well as the points of X because C contains $C_i^*(Y)$. By (a), $A_i = C_i^*(X)$, i.e., every member of $C_i^*(X)$ has an extension in C, and in order to prove that $Y = \beta_i X$ it remains only to show that C is contained in $C_i^*(Y)$: The binary relation $\leq'$ for Y defined by

$x \leq' y$ iff $f(x) \leq f(y)$ for all $f \in C$

is clearly a closed preorder for Y which is contained in the order of Y. It follows that $\leq'$ is an order. If $x, y \in X$ and $x \leq y$, then $f(x) \leq f(y)$ for all $f \in C$, i.e., $x \leq' y$. By 1.4(4) $\leq'$ is the order of Y and therefore the members of C, being $\leq'$ increasing, belong to $C_i^*(Y)$.

(b) implies (c). Assume that S is a convex subset of $\beta_i X$ which is contained in $\beta_i X \sim X$, satisfies

$$x \leq y \quad \text{whenever} \quad x, y \in X \quad \text{and} \quad x \in d(S), \; y \in i(S),$$

and has at least two elements. We may assume in addition that S is closed because the convex hull of any two-point subset of S has the same properties as S and is closed. Denote by Y the topological quotient space obtained from $\beta_i X$ by identifying the points of S, and denote by q the quotient map. It is well known that $(Y, q|X)$ is a topological compactification of X, i.e., Y is a compact Hausdorff topological space, $q|X$ is a topological embedding of X in Y, and $q[X]$ is dense in Y. Since Y and $\beta_i X$ are topologically distinct, we need only find a closed order for Y which turns it into a Nachbin compactification of X. From the proof of 2.2 (a) implies (b) we know not only that there is at most one such order but also where to look for it.

Set $C = \{f \in C^*(Y): f \circ q \in C_i^*(\beta_i X)\}$ and define a binary relation $\leq'$ for Y by

$$x \leq' y \quad \text{iff} \quad f(x) \leq f(y) \quad \text{for all} \quad f \in C.$$

$\leq'$ is clearly a closed preorder for Y. That it is an order with respect to which $q|X$ is an order embedding, will follow from the observation that

$$\{f \circ q: f \in C\} = \{g \in C_i^*(\beta_i X): g \text{ is constant on } S\}$$

combined with (i) and (ii) below.

(i) Let $x \in \beta_i X \sim S$. Since S is convex, either $x \notin d(S)$ or $x \notin i(S)$. If $x \notin d(S)$, the function $g: S \cup \{x\} \to \mathbb{R}$ defined by $g[S] = \{0\}$ and $g(x) = 1$ is continuous and increasing (with respect to the topology and order induced by $\beta_i X$) and so (see [4; pp. 48-49]) has an extension in $C_i^*(\beta_i X)$. Likewise, if $x \notin i(S)$, there exists a $g \in C_i^*(\beta_i X)$ such that $g[S] = \{1\}$ and $g(x) = 0$.

(ii) Let $x, y \in \beta_i X \sim S$ be distinct, say, $x \leq y$ is false.

As above, if $x \notin d(S)$, there exists a $g \in C_i^*(\beta_i X)$ such that $g[S] = \{0\}$, $g(x) = 1$, and $g(y) = 0$, and if $y \notin i(S)$, there exists a $g \in C_i^*(\beta_i X)$ such that $g[S] = \{1\}$, $g(x) = 1$, and $g(y) = 0$. Thus it remains to consider the case that $x \in d(S)$ and $y \in i(S)$. Then, by the hypotheses on S, not both x and y are in X and, moreover, $y \leq x$ is false. It follows that there exists a $g \in C_i^*(\beta_i X)$ such that $g[S] = \{0\}$, $g(x) = 0$, and $g(y) = 1$.

Now assume that $\leq$" is another closed order for Y with respect to which $q|X$ is an order embedding. We have to show that $\leq$" contains $\leq'$, and this is done as follows: If $x, y \in Y$ and $x \leq$" y is false, then there exists (Y with respect to $\leq$" is a compact ordered topological space) a $\leq$" increasing element f of $C^*(Y)$ such that $f(x) > f(y)$. Since $q|X$ is an order embedding with respect to $\leq$", $f \circ (q|X) \in C_i^*(X)$, i.e., $f \circ q \in C_i^*(\beta_i X)$ and so (see above) $f \in C$. Thus $x \leq' y$ is also false.

(c) implies (a). Assume that A is a closed subalgebra of $C^*(X)$ which contains the constant functions such that A_i separates points and closed subsets of X as well as the points of X and such that $A_i \neq C_i^*(X)$. Then $(A_i)\hat{} \neq C_i^*(\beta_i X)$ and so $c\ell[(A_i)\hat{} - (A_i)\hat{}] \neq C^*(\beta X)$. By the classical Stone-Weierstrass theorem there exist two distinct points $x_1, x_2 \in \beta_i X$ such that $f(x_1) = f(x_2)$ for all $f \in c\ell[(A_i)\hat{} - (A_i)\hat{}]$. Take S to be the convex hull of $\{x_1, x_2\}$. Obviously all members of $(A_i)\hat{}$ are constant on S. Since A_i separates points and closed subsets of X, S is contained in $\beta_i X \sim X$, and since A_i separates the points of X, S satisfies

$x \leq y$ whenever $x, y \in X$ and $x \in d(S)$, $y \in i(S)$.

This completes the proof of the theorem.

We note that the equivalence of (a) and (b) in the preceding theorem is a trivial consequence of an internal description of all Nachbin compactifications of a completely regular ordered topological space given in [8; Sec. 5].

2.3. COROLLARY (Hewitt [3; Theorems 4 and 5]). Let X be a completely regular Hausdorff topological space. The following conditions are equivalent.

(a) $A = C^*(X)$ whenever A is a closed subalgebra of $C^*(X)$ which contains the constant functions and separates points and closed subsets of X.

(b) X has a unique compactification.

(c) $\beta X \sim X$ has at most one element.

(d) Of any two disjoint zero sets of X one is compact.

Proof. (a), (b), and (c) are just the corresponding conditions of 2.2 for the case at hand, and so equivalent. The equivalence of (d) and (c) is a consequence of well known properties of βX.

2.4. COROLLARY ([5; Corollaries 2.9 and 2.11]). Let X be a completely regular ordered topological space whose order is total. The following conditions are equivalent.

(a) $A_i = C_i^*(X)$ whenever A is a closed subalgebra of $C^*(X)$ which contains the constant functions such that A_i separates points and closed subsets of X.

(b) X has a unique Nachbin compactification.

(c) A convex subset of $\beta_i X$ which is contained in $\beta_i X \sim X$ has at most one element.

(d) X is *order complete*, i.e., every nonempty subset of X which has an upper bound has a supremum.

Proof. (a), (b), and (c) are just the corresponding conditions of 2.2 for the case at hand, and so equivalent. The equivalence of (d) and (c) is a consequence of the description of $\beta_i X$ given in [5].

2.5. THEOREM. Let X be a completely regular ordered topological space. The following conditions are equivalent.

(a) $A_i = C_i^*(X)$ whenever A is a closed subalgebra of $C^*(X)$ which contains the constant functions such that A_i separates the points of X.

(b) A convex subset S of $\beta_i X$ which contains at most one point of X and satisfies

$$x \leq y \quad \text{whenever} \quad x, y \in X \quad \text{and} \quad x \in d(S), \; y \in i(S)$$

has at most one element.

Proof. (a) implies (b). Assume that S is a convex subset of $\beta_i X$ which contains at most one point of X, satisfies

$$x \leq y \quad \text{whenever} \quad x, y \in X \quad \text{and} \quad x \in d(S), \; y \in i(S),$$

and has at least two elements. We may assume in addition that S is closed because the convex hull of any two-point subset of S has the same properties as S and is closed. Setting

$$A = \{f|X: f \in C^*(\beta_i X) \text{ and } f \text{ constant on } S\},$$

we have

$$A_i = \{f|X: f \in C_i^*(\beta_i X) \text{ and } f \text{ constant on } S\},$$

and so $A_i \neq C_i^*(X)$. It remains to show that A_i separates the points of X. Let $x, y \in X$ be such that $x \leq y$ is false. Then either $x \notin d(S)$ or $y \notin i(S)$. The extension device already employed in the proof of 2.2 (b) implies (c) yields in the former case an $f \in C_i^*(\beta_i X)$ such that $f[S] = \{0\}$, $f(x) = 1$, and $f(y) = 0$, and in the latter an $f \in C_i^*(\beta_i X)$ such that $f[S] = \{1\}$, $f(x) = 1$, and $f(y) = 0$. So much for that.

(b) implies (a). Assume that A is a closed subalgebra of $C^*(X)$ which contains the constant functions such that A_i separates the points of X and $A_i \neq C_i^*(X) \cdot (A_i)\hat{\,} \neq C_i^*(\beta_i X)$ and so $c\ell[(A_i)\hat{\,} - (A_i)\hat{\,}] \neq C^*(\beta_i X)$. By the classical Stone-Weierstrass theorem there exist distinct $x_1, x_2 \in \beta_i X$ such that $f(x_1) = f(x_2)$ for all $f \in c\ell[(A_i)\hat{\,} - (A_i)\hat{\,}]$. Take S to be the convex hull of $\{x_1, x_2\}$. Obviously all members of $(A_i)\hat{\,}$ are constant on S. Since A_i separates the points of X, S contains at most one point of X and satisfies

$$x \leq y \quad \text{whenever} \quad x, y \in X \quad \text{and} \quad x \in d(S), \; y \in i(S).$$

This completes the proof of the theorem.

We note that the hypotheses in (b) of the preceding theorem are slightly redundant in that if S is any subset of $\beta_i X$ which satisfies

$$x \leq y \quad \text{whenever} \quad x, y \in X \quad \text{and} \quad x \in d(S), \; y \in i(S),$$

then the convex hull of S still satisfies this condition and contains at most one point of X.

2.6. COROLLARY (Hewitt [3; Theorem 3]). Let X be a completely regular Hausdorff topological space. The following conditions are equivalent.

(a) $A = C^*(X)$ whenever A is a closed subalgebra of $C^*(X)$ which contains the constant functions and separates the points of X.

(b) X is compact.

Proof. (a) and (b) are just the corresponding conditions of 2.5 for the case at hand.

2.7. COROLLARY. Let X be a completely regular ordered topological space whose order is total. The following conditions are equivalent.

(a) $A_i = C_i^*(X)$ whenever A is a closed subalgebra of $C^*(X)$ which contains the constant functions such that A_i separates the points of X.

(b) X is order complete and its topology is the order topology.

Proof. (a) and (b) are just the corresponding conditions of 2.5 for the case at hand (for (b) this is a consequence of the description of $\beta_i X$ given in [5]).

We conclude with a simple but typical example. Consider the set $X = \{0, 1\} \times [0, 1[$. Define two binary relations $\leq_1$ and $\leq_2$ for X by

$(n,t) \leq_1 (n',t')$ iff either $n < n'$ or $n = n'$ and $t \leq t'$,

$(n,t) \leq_2 (n',t')$ iff $n = n'$ and $t \leq t'$.

$\leq_1$ is a total order for X and $\leq_2$ is an order for X which is properly contained in $\leq_1$ but induces the same order as $\leq_1$ on $\{0\} \times [0, 1[$ and on $\{1\} \times [0, 1[$. Denote by τ_1 the $\leq_1$ order topology for X and by τ_2 the smallest topology for X which contains τ_1 and $\{1\} \times [0, 1[$. τ_2 properly contains τ_1 but induces the same topology as τ_1 on $\{0\} \times [0, 1[$ and on $\{1\} \times [0, 1[$.

(i) $(X, \tau_1, \leq_1)$ is a completely regular ordered topological space which satisfies the equivalent conditions of 2.5.

(ii) $(X, \tau_2, \leq_1)$ is a completely regular ordered topological space which satisfies the equivalent conditions of 2.2 but not those of 2.5.

(iii) $(X, \tau_2, \leq_2)$ is a completely regular ordered topological space which does not satisfy the equivalent conditions of 2.2.

The various implications of (i), (ii), and (iii) are best seen from the following illustration where ● indicates the points (0, 0) and (0, 1) of X, ⟶ the order of X, and o points in a Nachbin compactification of X which are not in X.

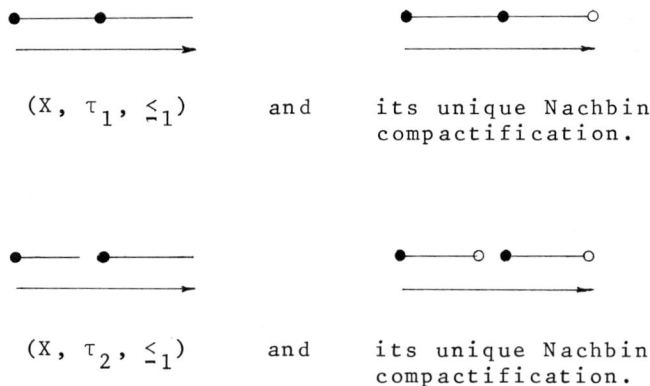

$(X, \tau_1, \leq_1)$ and   its unique Nachbin compactification.

$(X, \tau_2, \leq_1)$ and   its unique Nachbin compactification.

$(X, \tau_2, \leq_2)$ and its two Nachbin compactifications.

REFERENCES

1. M. H. Stone, Applications of the theory of Boolean rings to general topology, *Trans. Amer. Math. Soc.*, 41 (1937), 375-481.

2. K. Weierstrass, Über die analytische Darstellbarkeit sogenannter willkürlicher Funktionen reeller Argumente, *Sitzungsberichte Königl. Preuss. Akad. Wiss. Berlin*, 1885, pp. 633-639, 789-805. (*Mathematische Werke von Karl Weierstrass*, Mayer und Müller, Berlin, 1903, Band III, pp. 1-37.)

3. E. Hewitt, Certain generalizations of the Weierstrass approximation theorem, *Duke Math. J.*, 14 (1947), 419-427.

4. L. Nachbin, *Topology and Order*, Van Nostrand Mathematical Studies No. 4, D. Van Nostrand, Princeton, 1965.

5. J. Blatter, Order compactifications of totally ordered topological spaces, *J. Approximation Theory*, 13 (1975), 56-65.

6. J. Blatter and G. L. Seever, Interposition of semi-continuous functions by continuous functions, *Analyse fonctionelle et applications* (Comptes Rendus du Colloque d'Analyse, Rio de Janeiro, 1972), pp. 27-51, Actualités Sci. Indust., Hermann, Paris, 1974.

7. J. Blatter and G. L. Seever, Quasi-proximities and order compactifications, *Notices Amer. Math. Soc.*, Abstract 73 T - G 122, 20 (1973), A 594.

8. J. Blatter and G. L. Seever, Interposition and lattice cones of functions, *Trans. Amer. Math. Soc.* (to appear).

9. G. Hommel, *Isotone Radon Masse auf lokalkompakten geordneten Räumen*, Dissertation, Universität Erlangen Nürnberg, 1973.

CLASSES OF DISTRIBUTIONS WITH PARABOLIC MAXIMAL FUNCTIONS IN $L^p(\mathbb{R}^n)$*

Alberto P. Calderón[†]

Department of Mathematics
Massachusetts Institute of Technology
Cambridge, Massachusetts

and

Alberto Torchinsky[‡]

Department of Mathematics
Cornell University
Ithaca, New York

This chapter describes an extension of some interesting recent results of Burkholder et al. [1] and Fefferman and Stein [2] concerning the classes H^p of analytic functions in a half-space. Complete proofs of the results will appear elsewhere (Calderón and Torchinsky [3]).

We study functions $F(x, t)$ defined for $x \in \mathbb{R}^n$, $t > 0$, and in greater detail, convolutions of the form $F(x, t) = (f * \phi_t)(x)$, where $f \in S'$ is a tempered distribution in $\mathbb{R}^n$ and

---

[*]Research partly supported by GP 36775.

[†]Present address: Department of Mathematics, University of Chicago, Chicago, Illinois.

[‡]Present address: Department of Mathematics, Indiana University, Bloomington, Indiana.

ϕ_t is an approximate identity formed with "parabolic" dilations of a function ϕ in the L. Schwartz class S of infinitely differentiable, rapidly decreasing functions in $\mathbb{R}^n$. The justification for the study of these functions lies in the fact they are closely related to the study of the so-called singular integrals with mixed homogeneity and of related multiplier and differential operators. With such functions $F(x, t)$ we associate maximal functions and analogs of the Lusin and Littlewood-Paley functions, and study the relations among these for a given f and various choices of ϕ. We also introduce analogs of the H^p classes and extend to this case the theory of Fefferman and Stein. The results are the same, and although for the sake of simplicity we have restricted the functions ϕ to the class S, there is no difficulty in generalizing the theory to larger classes. Methodologically, however, the present situation is different, since the theory of harmonic functions is not available as a tool. Yet some of the new techniques developed by Burkholder et al. and Fefferman and Stein can still be used in our case.

We let $\{A_t\}_{t>0}$, $A_{ts} = A_t A_s$ be a continuous group of affine transformations of $\mathbb{R}^n$ leaving the origin fixed, and denote its infinitesimal generator by P so that $t(d/dt)A_t = PA_t$. We also assume that for some α, β with $1 \leq \alpha \leq \beta$ we have $t^\alpha |x| \leq |A_t x| \leq t^\beta |x|$, $x \in \mathbb{R}^n$ and $t \geq 1$.

The transposed transformations A_t^* of A_t with respect to the usual inner product in $\mathbb{R}^n$ also form a multiplicative group and satisfy similar inequalities. As is well known, $\det A_t = \det A_t^* = t^\gamma$, $\gamma = $ trace of P. With T_t we will denote the operator defined by $(T_t f)(x) = f(A_t x)$.

Given a function $\varphi \in S$ we define the "parabolic" dilations alluded to above as $\phi_t(x) = t^{-\gamma} \phi(A_t^{-1} x) = t^{-\gamma}(T_t^{-1}\phi)(x)$. A special role, somehow analogous to that of the Poisson kernel in the case of harmonic functions, is played by $\phi_t(x)$ with $\hat{\phi}(x) = \exp[-4\pi^2 (Bx, x)]$ where B is the unique strictly positive, self-adjoint matrix such that $PB + BP^* = I$, I being the identity transformation of $\mathbb{R}^n$. This particular function satisfies a differential equation, as is readily seen by taking Fourier transforms, namely

PARABOLIC MAXIMAL FUNCTIONS   29

$$t \frac{\partial}{\partial t} \phi_t(x) = T_t^{-1} \Delta T_t \phi_t(x),$$

where Δ is the Laplacian.

   With each such group A_t we associate a translation invariant distance function $r(x)$ defined as follows (de Guzmán [4]): let $r(x)$ be the unique value of t such that $|A_t^{-1} x| = 1$, $r(0) = 0$. Then $r(x + y) \leq r(x) + r(y)$; $r(A_t x) = t r(x)$, and $|x|$ lies between $r(x)^\alpha$ and $r(x)^\beta$. Furthermore, $r(x) \in C^\infty(\mathbb{R}^n - \{0\})$, and if D is a monomial differential operator with constant coefficient of order m, then $|D r(x)| \leq c \max[r(x)^{1-\alpha m}, r(x)^{1-\beta m}]$. Similarly for $r^*(x)$ defined with A_t^* in place of A_t. In what follows, c will denote a constant depending only on the parameters and the kernels ϕ, ψ, ... etc. involved.

   Corresponding to the metric $r(x)$ we have the following maximal theorem.

THEOREM. Let $|\psi(x)| \leq \eta[r(x)]$, where $\eta(t)$ is strictly decreasing in $0 \leq t < \infty$ and $\int \eta[r(x)] \, dx = 1$. Then if $F(x, t) = (f * \psi_t)(x)$ and $M_a(x) = \sup_{r(y) \leq at} |F(x - y, t)|$, $a \geq 0$, we have $\|M_a\|_p \leq C \|f\|_p$, $1 < p \leq \infty$, and $|\{M_a(x) > s\}| \leq c \|f\|_1 / s$, where $|\cdot|$ denote Lebesgue measure in $\mathbb{R}^n$.

   The theorem then holds if $|\psi(x)| \leq C(1 + |x|)^{-m}$ for sufficiently large m ($> \gamma/\alpha$). Also, if $a = 0$ and $\psi(x) = \chi(x)$, χ being the characteristic function of the unit ball, then the maximal function obtained is the analog of the usual Hardy-Littlewood maximal function and is denoted by $h(x, f)$.

   We now describe some general facts about functions $F(x, t)$, $x \in \mathbb{R}^n$, $t > 0$. We define the maximal functions

$$M_a(x, F) = M_a(x) = \sup_{y, t} |F(y, t)| \chi\left\{\frac{r(x - y)}{at}\right\}$$

$$= \sup_{r(y) \leq at} |F(x - y, t)|, \quad (a \geq 0),$$

and

$$N_\lambda(x, F) = N_\lambda(x) = \sup_{y,t} |F(y, t)| \left(1 + \frac{r(x - y)}{t}\right)^{-\lambda} \quad (\lambda \geq 0).$$

To illustrate some of the methods we employ, we prove the following.

THEOREM 1. Let $a, b, \lambda > 0$. Then for $s > 0$ we have

$$|\{M_a(x) > s\}| \leq c\left(1 + \frac{a}{b}\right)^\gamma |\{M_b(x) > s\}|,$$

and consequently $\|M_a\|_p \leq c\|M_b\|_p$, $0 < p < \infty$. Moreover, if $M_b \in L^p$, then $\|N_\lambda\|_p \leq c\|M_b\|_p$ for $\lambda > \gamma/p$.

Proof. Let

$$O = \{M_b(x) > s\} \quad \text{and} \quad O_1 = \left\{h(x, 1_O) \geq \left(\frac{b}{a+b}\right)^\gamma\right\},$$

where 1_O is the characteristic function of O. Then by the maximal theorem

$$|O_1| \leq c(\gamma)\left(1 + \frac{b}{a}\right)^\gamma |O|.$$

We will now show that $\{M_a(x) > s\} = O_2 \subset O_1$. In fact, if $x \in O_2$, then there is (y, t) with $r(x - y) \leq at$ and $|F(y, t)| > s$, and consequently $B(y, bt) = \{z: r(y - z) \leq bt\} \subset O$. Also $B(x, at + bt) = \{z: r(x - z) \leq at + bt\} \supset B(y, bt)$. Thus for $x \in O_2$ we have

$$h(x, 1_O) \geq \frac{|O \cap B(x, at + bt)|}{|B(x, at + bt)|} \geq \frac{|B(y, bt) \cap B(x, at + bt)|}{|B(x, at + bt)|}$$

$$= \frac{|B(y, bt)|}{|B(x, at + bt)|} = \left(\frac{b}{a+b}\right)^\gamma, \quad \text{and} \quad x \in O_1.$$

The pth-norm inequalities for the M functions now follow from the fact that $\|g\|_p^p = p\int_0^\infty s^{p-1}|\{g > s\}|\,ds$.

Also, as is readily verified, we have

$$(1 + s)^{-\lambda} \le \sum_{k=1}^{\infty} 2^{-k} \chi(s/2^{k/\lambda}),$$

where χ is the characteristic function of the interval $[0, 1]$. Thus we have

$$N_\lambda(x) \le \sum_{k=1}^{\infty} 2^{-k} \sup_{y,t} |F(y, t)| \chi\left(\frac{r(x - y)}{t 2^{k/\lambda}}\right)$$

$$= \sum_{k=1}^{\infty} 2^{-k} M_{2^{k/\lambda}}(x).$$

Now if $0 < p \le 1$ we have

$$N_\lambda^p(x) \le \sum_{k=1}^{\infty} 2^{-pk} M_{2^{k/\lambda}}^p(x),$$

and integrating we obtain $\|N_\lambda\|_p^p \le C \|M_b\|_p^p$ provided that $\lambda > \gamma/p$. For $p > 1$ we obtain the same result by means of Minkowski's inequality. The same restriction on λ is readily seen to imply the convergence of the sum of the norms.

THEOREM 2. (Hardy and Littlewood [5]) Suppose that $M_a(x, F) \in L^p$, $a > 0$, $0 < p < \infty$. Then for $p < q < \infty$

$$\left(\int\int |F(x, t)|^q t^{\gamma(q/p-1)} \frac{dx\, dt}{t}\right)^{1/q} \le C \|M_a\|_p,$$

and

$$\sup_{t>0} (|F(x, t)| t^{\gamma/p}) \le C \|M_a\|_p \quad (q = \infty).$$

THEOREM 3. Let $F(x, t)$ be continuously differentiable with respect to x in $t > 0$. Let

$$H(x, t) = \sum \left|\left(T_t^{-1} \frac{\partial}{\partial x_i} T_t F\right)(x, t)\right|$$

and suppose that $M_a(x, F)$ and $M_b(x, H)$, $a, b > 0$ are in L^p,

$0 < p < \infty$. Then there is a constant $c = c(a, b, p)$ such that

$$\|M_a(x, F)\|_p \leq c \|M_0(x, F)\|_p^{p/(p+\gamma)} \|M_b(x, H)\|_p^{\gamma/(p+\gamma)}$$

if $\|M_0(x, F)\|_p \leq \|M_b(x, H)\|_p$, and $\|M_a(x, F)\|_p \leq c \|M_0(x, F)\|_p$ otherwise.

DEFINITION. With $F(x, t)$ defined for $x \in \mathbb{R}^n$, $t > 0$, we associate the Lusin Function $S_a(x, F)$, or briefly $S_a(x)$,

$$S_a^2(x) = \frac{1}{\omega} \int |F(y, t)|^2 \chi\left(\frac{r(x - y)}{at}\right) (at)^{-\gamma} dy \frac{dt}{t} \quad (a > 0)$$

where ω is the volume of the unit ball, and the Littlewood-Paley function $G_\lambda(x, F)$, or briefly $G_\lambda(x)$,

$$G_\lambda^2(x) = \int |F(y, t)|^2 \left(1 + \frac{r(x - y)}{t}\right)^{-2\lambda} dy \frac{dt}{t} \quad (\lambda > 0).$$

Except for some technical differences one can show the following result by methods similar to those used in proving Theorem 1.

THEOREM 4. Let $F(x, t)$ be defined for $x \in \mathbb{R}^n$, $t > 0$, and for $a, b, \lambda > 0$ consider S_a, S_b, and G_λ. Suppose that $S_b \in L^p$, $0 < p \leq 2$; then

$$\|S_a\|_p^p \leq \frac{c}{2 - p} \left(1 + \frac{a}{b}\right)^{\gamma(1-2/p)} \|S_b\|_p^p \quad (p < 2),$$

and $\|S_a\|_2 = \|S_b\|_2$. Moreover $\|G_\lambda\|_p \leq c \|S_b\|_p$, $0 < p \leq 2$, and $\lambda > \gamma/p$.

We now consider $F(x, t) = (f * \phi_t)(x)$, $f \in S'$, $\phi \in S$. We consider relations between the maximal, Lusin, and Littlewood-Paley functions associated with $F(x, t)$ for various choices of ϕ. The main tool in passing from one ϕ to another

PARABOLIC MAXIMAL FUNCTIONS

is a generalization of a representation lemma proved for radial functions in [6].

DEFINITION. We say that a family $\{\phi_j\}_{j=1}^m \subset S$ is *nonvanishing* if $\sum_{j=1}^m |\hat{\phi}_j(A_t^* x)|^2$ does not vanish identically in $t > 0$ for $x \neq 0$.

REPRESENTATION LEMMA. Suppose $\{\phi_j\}_1^m \subset S$ is nonvanishing. Then there exists a family $\{\eta_j\}_1^m \subset S$ such that $\hat{\eta}_j \in C_0^\infty(\mathbb{R}^n)$, the support of the $\hat{\eta}_j$ is disjoint from the origin, and for each compact set K not containing the origin there exists $\delta > 0$ such that

$$\int_\varepsilon^{1/\varepsilon} \sum_{j=1}^m \hat{\eta}_j(A_t^* x)\, \hat{\phi}_j(A_t^* x)\, \frac{dt}{t} = 1$$

for $x \in K$ and $\delta > \varepsilon > 0$.

THEOREM 5. Let $f \in S'$ and $\{\phi^{(j)}\}_1^m \subset S$ be nonvanishing. Suppose that for $F_j(x, t) = (f * \phi_t^{(j)})(x)$ we have $M_a(x, F_j) \in L^p(\mathbb{R}^n)$, $0 < p \leq 2$, $a > 0$. Then there is a function $h(x)$ locally integrable in $\mathbb{R}^n - \{0\}$ such that $\hat{f} - h(x)$ is a distribution supported at the origin. Furthermore, if $p < q/(q - 1) \leq 2$, then

(1) $$\int |h(x)|^q\, r^*(x)^{-\gamma(q/p-q+1)}\, dx \leq c\left(\sum_1^m \|M_a(\cdot, F_j)\|_p\right)^q$$

and if $p < 1$, then $h(x)$ is continuous and

(2) $$|h(x)| \leq c r^*(x)^{\gamma(1/p-1)} \sum_1^m \|M_a(\cdot, F_j)\|_p.$$

If the set $\{\phi^{(j)}\}$ reduces to a single function ϕ with nonvanishing integral, then $\hat{f} = h(x)$ and (1) holds for $p \leq q/(q - 1) \leq 2$ and (2) for $p \leq 1$.

THEOREM 6. Let f, $\{\phi^{(j)}\}$, $F_j(x, t)$ be as in Theorem 5. Suppose that for sufficiently large k, $\hat{f}(x)(1 + |x|^2)^{-k}$ coincides with an integrable function. Let $\psi \in S$, $\hat{\psi}(0) = 0$, and set $H(x, t) = (f * \psi_t)(x)$. Then $M_a(x, H) \leq \sum_{j=1}^{m} N_\lambda(x, F_j)$, $\lambda > 0$. If $\{\phi^{(j)}\}$ reduces to a single function ϕ with $\hat{\phi}(0) \neq 0$, then result holds for all $\psi \in S$. Furthermore the theorem is valid if f coincides with a function in L^p, $1 \leq p \leq \infty$.

Proof. We prove the first part of the theorem. In view of the assumptions $\hat{f}$ coincides with a locally integrable function $h(x)$, and $h(x)\hat{\psi}(A_t^* x)$ is integrable for all $t > 0$. Consequently $H(x, t) = \int \exp[-2\pi i(x,y)] h(y) \hat{\psi}(A_t^* y)\, dy$, and the integral is absolutely convergent. By the representation lemma we have

$$H(x, t) = \lim_{\varepsilon \downarrow 0} \sum_{j=1}^{m} \iint_{\varepsilon}^{1/\varepsilon} e^{-2\pi i(x,y)} h(y) \hat{\psi}(A_t^* y) \hat{\eta}^{(j)}(A_s^* y)$$

$$\times \hat{\phi}^{(j)}(A_s^* y) \frac{ds}{s}\, dy.$$

Since these integrals are absolutely convergent we may change the order of integration and replacing $\hat{\eta}^{(j)}(A_s^* y) \hat{\psi}(A_t^* y)$ by $\chi_{s,t}^{(j)}(y)$, that is $\eta_s^{(j)} * \psi_t = \chi_{s,t}^{(j)}$, we obtain

$$H(x, t) = \lim_{\varepsilon \to 0} \sum_{j=1}^{m} \int_{\varepsilon}^{1/\varepsilon} \int F_j(x - y, s) \chi_{s,t}^{(j)}(y)\, dy\, \frac{ds}{s},$$

whence

$$|H(x, t)| \leq \sum_{j=1}^{m} \int |F_j(x - y, s)| |\chi_{s,t}^{(j)}(y)|\, dy\, \frac{ds}{s},$$

and

$$|H(x + w, t)| \leq \sum_{j=1}^{m} \sup_{y,s} \left\{ |F_j(x - y, s)| \left(1 + \frac{r(y)}{s}\right)^{-\lambda} \right\}$$

$$\times \int |\chi_{s,t}^{(j)}(y + w)| \left(1 + \frac{r(y)}{s}\right)^{\lambda} dy\, \frac{ds}{s}.$$

Now, one can show that the last integral is finite and bounded for $r(w) \leq t$. Thus taking the supremum over the set $r(w) \leq t$ we obtain

$$M_1(x, H) \leq c \sum_{j=1}^{m} N_\lambda(x, F_j).$$

The same result holds for values of a other than 1 by observing that if $\tilde{\psi}(x) = a^\gamma \psi(A_a x)$ and $\tilde{H}(x, t) = (f * \tilde{\psi}_t)(x)$, then $\tilde{H}(x, t) = H(x, t/a)$ and $M_a(x, H) = M_1(x, \tilde{H})$. If the $\phi^{(j)}$'s reduce to a single ϕ with $\hat{\phi}(0) \neq 0$, then the assertion certainly holds for $\psi_1(x) = \hat{\psi}(0) \hat{\phi}(0)^{-1} \phi(x) - \psi(x)$, and consequently also for $\psi(x)$. The remaining parts are proved in a similar fashion.

THEOREM 7. Let $f \in S'$, $\phi, \psi \in S$, $\hat{\phi}(0) \neq 0$. Set $F(x, t) = (f * \phi_t)(x)$ and $H(x, t) = (f * \psi_t)(x)$. Then if $M_a(x, F) \in L^p$, $0 < p \leq \infty$, $a > 0$, then $M_a(x, H) \in L^p$ and $\|M_a(\cdot, H)\|_p \leq c\|M_a(\cdot, F)\|_p$ and $\|M_a(\cdot, F)\|_p \leq c\|M_0(\cdot, F)\|_p$.

THEOREM 8. Let $f \in S'$, $\phi \in S$ with $\hat{\phi}(0) \neq 0$. If $M_a(x, F) \in L^p$, $1 \leq p \leq \infty$, then f is a function in L^p and $|f(x)||\int \phi(y) \, dy| \leq M_a(x)$.

THEOREM 9. Let $f \in S'$, $\{\phi^{(j)}\}_1^m \subset S$ be nonvanishing. Set $F^{(j)}(x, t) = (f * \phi_t^{(j)})(x)$, and for $\psi \in S$ with $\hat{\psi} \in C_0^\infty$, supp $\hat{\psi}$ disjoint from the origin, set $H(x, t) = (f * \psi_t)(x)$. We then have $M_a(x, H) \leq c \sum_{j=1}^{m} G_\lambda(x, F_j)$. Moreover, if $S_a(x, F_j) \in L^p$, $0 < p \leq 2$, $a > 0$, then $M_a(x, H) \in L^p$ and $\|M_a(\cdot, H)\|_p \leq c \sum_1^m \|S_a(\cdot, F_j)\|_p$.

THEOREM 10 (multiplier theorem). Let $f \in S'$ such that $\hat{f} = h(x)$, with $|h(x)|^2 (1 + |x|)^{-\ell}$ integrable for sufficiently large ℓ. Let $\phi, \eta \in S$, $\hat{\phi}(0) = 0$, $\hat{\psi} = \hat{\phi} \hat{\eta}$. Let $k(x)$ be a bounded function such that the inverse Fourier transform $g(x, t)$ of $\hat{\eta}(x) k(A_t^* x)$ has the property that $g(x, t)[1 + r(x)]^\lambda \in L^2$ for each t and its norm is bounded by c. Let $\hat{f}_1(x) = h(x) k(x)$, $F(x, t) = (f * \phi_t)(x)$, $H(x, t) = (f_1 * \psi_t)(x)$. Then

$$|H(x + w, t)|^2 \le c\left(1 + \frac{r(w)}{t}\right)^{2\lambda} t^{-\gamma}\left[\int |F(x - y, t)|^2 \times \left(1 + \frac{r(y)}{t}\right)^{-2\lambda} dy\right],$$

where c is as above, and if $\mu - \lambda > \gamma/2$, then

$$G_\mu(x, H) \le cG_\lambda(x, F)\left\{\int [1 + r(y)]^{2(\lambda-\mu)} dy\right\}^{1/2}$$

where c is the same as above.

Suppose, on the other hand, that $f \in L^p$, $1 \le p \le \infty$, and that $\lambda > \gamma/2$. Let $g^{(t)}$ be the inverse Fourier transform of $\hat{\psi}(A_t^* x) k(x)$. Then $g^{(t)}$ is integrable, $H(x, t) = (f * g^{(t)})(x)$ is well defined and coincides with the function $H(x, t)$ above if f satisfies in addition to the preceding assumptions. Furthermore, $H(x, t)$ also satisfies the preceding inequalities.

Proof. Suppose first that $\hat{f} = h(x)$. Then, since $h(x) \hat{\phi}(A_t^* x)$ and $k(x) \hat{\eta}(A_t^* x)$ are square-integrable functions for each $t > 0$, and the Fourier transform $H(x, t)$ coincides with $h(x) \hat{\phi}(A_t^* x) k(x) \hat{\eta}(A_t^* x)$, we have

(3) $$H(x, t) = \int F(x - y, t) g_1(y, t) dy,$$

where $g_1(y, t)$ is the inverse Fourier transform of $\hat{k}(x) \hat{\eta}(A_t^* x)$. This expression for $H(x, t)$ also holds when $f \in L^p$, $1 \le p \le \infty$, and $\lambda > \gamma/2$. From (3) it follows that

$$|H(x + w, t)|^2 \le \left[\int |F(x - y, t)|^2\left(1 + \frac{r(y)}{t}\right)^{-2\lambda} dy\right] \times \left[\int |g_1(y + w, t)|^2\left(1 + \frac{r(y)}{t}\right)^{2\lambda} dy\right].$$

Setting $y = A_t \bar{y}$, $w = A_t \bar{w}$, the last integral becomes

$$\int |g_1(A_t(\bar{y} + \bar{w}), t)|^2 \, t^\gamma [1 + r(\bar{y})]^{2\lambda} \, d\bar{y}$$

$$\leq [1 + r(w)]^{2\lambda} \int |g_1(A_t(\bar{y} + \bar{w}), t)|^2 [1 + r(\bar{y} + \bar{w})]^{2\lambda} t^\gamma \, d\bar{y}.$$

But the Fourier transform of $g_1(A_t y, t) t^\gamma$ is precisely $k(A_t^{*-1} x) \hat{n}(x)$. Thus according to our assumptions, the last integral is majorized by $ct^{-\gamma}$, and substituting in (3), the first inequality of our theorem follows. Further, multiplying that inequality by $\{1 + [r(w)/t]\}^{-2\mu} t^{-\gamma}$ and integrating we obtain

$$\int |H(x + w, t)|^2 \left(1 + \frac{r(w)}{t}\right)^{-2\mu} t^{-\gamma} \, dw \, \frac{dt}{t}$$

$$\leq c \int |F(x - y, t)|^2 \left(1 + \frac{r(y)}{t}\right)^{-2\lambda} \left(1 + \frac{r(w)}{t}\right)^{2(\lambda-\mu)} t^{-2\gamma} \, dy \, dw \, \frac{dt}{t}$$

and since for $2(\lambda - \mu) < -\gamma$ the integral

$$\int \left(1 + \frac{r(w)}{t}\right)^{2(\lambda-\mu)} dw = t^\gamma \int [1 + r(w)]^{2(\lambda-\mu)} \, dw$$

is finite, integrating above first with respect to w, the desired result follows.

THEOREM 11. Let $f \in S'$ be such that $\hat{f} = h(x)$, where $h(x)$ is locally square integrable and $|h(x)|^2(1 + |x|)^{-\ell}$ is summable for sufficiently large ℓ. Let $\{\phi^{(j)}\}_1^m \subset S$ be non-vanishing and $\phi^{(j)}(0) = 0$, $1 \leq j \leq m$. Set $F_j(x, t) = (f * \phi_t^{(j)})(x)$ and suppose that $\hat{\psi} \in S$ is such that $\hat{\psi}(0) = 0$. Let $H(x, t) = (f * \psi_t)(x)$. Then if $\mu - \lambda > \gamma$,

$$G_\mu(x, H) \leq c \sum_{j=1}^m G_\lambda(x, F_j).$$

The same conclusion holds if $f \in L^p$, $1 \leq p \leq \infty$, and $\lambda > \gamma/2$.

THEOREM 12. Let $f \in S'$, $\phi, \psi \in S$, $\hat{\phi}(0) \neq 0$, $\hat{\psi}(0) = 0$. Set $F(x, t) = (f * \phi_t)(x)$ and $H(x, t) = (f * \psi_t)(x)$. Suppose that $M_1(x, F) \in L^p$, $0 < p < \infty$. Then $G_\lambda(x, H) \in L^p$ and $\|G_\lambda(\cdot, H)\|_p \leq c \|M_1(\cdot, F)\|_p$ provided that $\lambda > \gamma/p$ if $0 < p \leq 2$, and $\lambda > \gamma[(3/2) - (2/p)]$ if $p > 2$.

THEOREM 13. Let $\{\psi^{(j)}\}_1^m \subset S$ be nonvanishing and $\hat{\psi}^{(j)}(0) = 0$, and $\phi \in S$. For $f \in S'$ such that $\hat{f}(1 + |x|^2)^{-\ell} \in L^2$ for sufficiently large ℓ set $F(x, t) = (f * \phi_t)(x)$, $K_j(x, t) = (f * \psi_t^{(j)})(x)$, $1 \leq j \leq m$. Then if $S_1(x, K_j) \in L^p$, $1 \leq j \leq m$, $0 < p < \infty$, we have

$$M_1(x, F) \in L^p \quad \text{and} \quad \|M_1(\cdot, F)\|_p \leq C \sum_{j=1}^m \|S_1(\cdot, K_j)\|_p.$$

REFERENCES

1. D. L. Burkholder, R. F. Gundy, and M. L. Silverstein, *Trans. Amer. Math. Soc.*, 157 (1971), 137-153.
2. C. Fefferman and E. M. Stein, *Acta Math.*, 129 (1972), 137-193.
3. A. P. Calderón and A. Torchinsky, *Advan. Math.*, to appear.
4. M. de Guzmán, *Rev. Acad. Sci. (Spain)*, 44 (1970), 77-137.
5. G. H. Hardy and J. E. Littlewood, *Math. Z.*, 34 (1932), 403-439.
6. A. P. Calderón, *Studia Math.*, 24 (1964), 113-190.
7. A. P. Calderón, *Studia Math.*, 44 (1972), 563-582.
8. P. L. Duren, B. W. Romberg, and A. L. Shields, *J. Reine Ang. Math.*, 238 (1969), 32-60.

LOCALLY SOLVABLE PSEUDODIFFERENTIAL OPERATORS WITH DOUBLE CHARACTERISTICS

Fernando Cardoso

Instituto de Matemática
Universidade Federal de Pernambuco
Recife, Pernambuco, Brazil

and

The Institute for Advanced Study
Princeton, New Jersey

We shall describe here work done in collaboration with F. Treves. Details of the proof will appear in [1]. We denote by

$$\mathbb{P} = P(x, D) \sim \sum_{j=0}^{+\infty} P_{m-j}(x, D),$$

a pseudodifferential operator of order m, say in an open subset Ω of a Euclidean space $\mathbb{R}^N$, $N \geq 2$. We shall argue under the hypothesis that the principal symbol $P_m(x, \xi)$ of $\mathbb{P}$ can be factorized as follows:

$$P_m(x, \xi) = Q(x, \xi)[L(x, \xi)]^2 \qquad (1)$$

in a conic neighborhood U of a point (x_0, ξ^0) of $\dot{T}^*(\Omega)$, the complement of the zero section in the cotangent bundle $T^*(\Omega)$ over Ω. The factors Q, L are C^∞ functions in U, positive-homogeneous of degree $m - 2$ and 1 respectively with respect to ξ, and have the following properties:

$$L(x_0, \xi^0) = 0, \quad d_\xi L(x_0, \xi^0) \neq 0, \quad Q(x_0, \xi^0) \neq 0. \qquad (2)$$

We shall denote by $\Gamma_z(x_0, \xi^0)$ the *bicharacteristic strip* of $\mathrm{Re}(zL(x, \xi))$ through (x_0, ξ^0). This presumes that $d_\xi \mathrm{Re}(zL)$ does not vanish at (x_0, ξ^0); z is a complex number $\neq 0$. Let us regard momentarily $L(x, \xi)$, suitably extended to the complement of U, as the symbol of a first-order pseudodifferential operator $L(x, D)$ of *principal type*. We recall the following conditions often used in the solvability theory of operators of principal type:

(FZ): Whatever $(x_0, \xi^0) \in \dot{T}^*(\Omega)$ where $L(x, \xi)$ vanishes, whatever $z \in \mathbb{C}$ such that $d_\xi \mathrm{Re}(zL) \neq 0$ at (x_0, ξ^0), the restriction of $\mathrm{Im}(zL)$ along $\Gamma_z(x_0, \xi^0)$ vanishes of finite order k at (x_0, ξ^0).

<u>Remark 1</u>. It can be shown (see [2]) that k is independent of all such z.

(ψ): Whatever $(x_0, \xi^0) \in \dot{T}^*(\Omega)$ and whatever complex number z, as above, if the restriction of $\mathrm{Im}(zL(x, \xi))$ to $\Gamma_z(x_0, \xi^0)$ is negative at some point, then it remains nonpositive from then on along $\Gamma_z(x_0, \xi^0)$ (bicharacteristic strips are oriented).

The following result is known.

THEOREM 1. Assuming that (FZ) is true, the principal type operator L is locally solvable in Ω if and only if (ψ) holds.

The necessary part of the theorem is due to Yu. V. Egorov and also to Nirenberg and Treves [2]. The sufficiency follows from a theorem announced by Egorov which states that the conjunction of (FZ) and (ψ) implies that $^t L$, the transpose of L, is subelliptic in Ω.

Returning to the operator $\mathbb{P}$ satisfying (1) and (2) [(x_0, ξ^0) is then referred to as a *double characteristic point of constant multiplicity* of $\mathbb{P}$] we proved in [1]:

THEOREM 2. Assume that (FZ) holds for the first-order factor L and that (ψ) is violated by L at (x_0, ξ^0). Then $\mathbb{P}$ is not locally solvable at x_0.

The noteworthy feature in Theorem 2 is that the lower-order terms of $\mathbb{P}$ do not influence the conclusion.

COROLLARY 1. Suppose that $\mathbb{P}$ is a differential operator in Ω and that (FZ) holds with an odd integer k. Then, neither $\mathbb{P}$ nor $^t\mathbb{P}$ is locally solvable at x_0.

When k = 1, Theorem 2 can be more easily proved and follows also from recent results of J. Sjöstrand [3].

The next case to investigate is the one where the factor L [still satisfying (FZ)] satisfies the condition (ψ) for local solvability. In such case obstruction to solvability may come from the Mizohata-Ohya invariant

$$\sigma(x, \xi) = P_{m-1}(x, \xi) - \frac{1}{2} \sum_{|\alpha|=1} \partial_\xi^\alpha D_x^\alpha P_m(x, \xi)$$

specifically from the fact that it does not decay to zero fast enough as (x, ξ) converges to (x_0, ξ^0) along certain curves in the cotangent bundle.

Although the Mizohata-Ohya invariant plays no role in the statement of Theorem 2, it is very much present in its proof, which subdivides into two parts, according to whether the lower-order terms have or do not have a "strong influence." The precise measurement of this influence is achieved by means of the imaginary part of the square root of the invariant in question, defined in the characteristic variety.

The starting point in the proof of Theorem 2 is the same as always in this kind of question: the remark of Hörmander as to the functional-analytic consequence of local solvability: if $\mathbb{P}$ were locally solvable at x_0 there would be two neighborhoods $V \subset U$ of x_0 in Ω, a compact subset K of U, an integer $M \geq 0$, a constant $C > 0$ such that

$$\left| \int fv \, dx \right| \leq C \sup \left(\sum_{|\alpha| \leq M} |D^\alpha f| \right) \cdot \sup_K \left(\sum_{|\alpha| \leq M} |D^\alpha ({}^t\mathbb{P} v)| \right) \qquad (3)$$

for every $f, v \in C_c^\infty(V)$. Here $\alpha = (\alpha_1, \ldots, \alpha_N)$ is a multi-index of nonnegative integers and $D^\alpha f = D_1^{\alpha_1} \cdots D_N^{\alpha_N} f$ with $D_j = -i\partial/\partial x_j$. The proof of Theorem 2 will consist in proving that, in the present situation, Eq. (3) cannot hold, whatever the choice of U, V, K, M, C. In order to show this one takes

$$v = e^{i\rho w}\varphi, \quad \rho \sim +\infty, \qquad (4)$$

with the complex-valued (phase) function $w \in C^\infty(\Omega)$ and the (amplitude) $\varphi \in C_c^\infty(V)$ chosen in such a way that v is an "approximate" solution of the homogeneous equation ${}^t\mathbb{P} v = 0$. An important role in the investigation of problems of the type with which we are concerned here is played by an asymptotic expansion in power of ρ, about $\rho \sim +\infty$, of $\mathbb{P}(e^{i\rho w}\varphi)$ (the gradient of w at x_0 is equal to ξ^0). Special, and generally cruder, versions of such an expansion are customarily used in pseudodifferential operator theory (see [2, 4, 5]). However, dealing with multiple characteristics demands a greater precision in the estimate of the remainder, in the asymptotic expansion, than was needed in the study of ψdo's of principal type, and although the techniques to establish such expansions are fairly standard, this entails a few modifications that are not completely self-evident. It should also be remarked that when applying the asymptotic expansion we need (since w is complex-valued) a "good" analytic approximation, in the covariable ξ, of the symbol ${}^t P(x, \xi)$ of ${}^t \mathbb{P}$. Later, when combining the asymptotic expansion with Eq. (3), we perform also an analytic approximation in the variable x. All of this, of course, introduces some "errors" which must be estimated when disproving Eq. (3).

We shall outline the proof of Theorem 2 [i.e., the impossibility of Eq. (3) to hold] for a very simple example from which most of the basic ideas of the general proof orginated and in which, fortunately, most of the technical difficulties of the latter are not present. Roughly speaking, one can "reduce" the general situation to one closer to the example by performing "admissible operations" such as: use of *canonical transformations* to straighten up bicharacteristics and to flatten transversal "pieces" of the characteristic variety perpendicular to the straightened bicharacteristics (see [6]); moving away (along the characteristic variety) from the original point (x_0, ξ^0) to a new point (x_0', ξ'^0), *arbitrarily* close to (x_0, ξ^0) and *conveniently* chosen, so that the problem of disproving solvability of $\mathbb{P}$ at x_0 is transferred to that of the nonsolvability of $\mathbb{P}$ at x_0' (in this step it is *very important* that the order k of the zero in Theorem 2 is necessarily *odd*); "division" by the elliptic factor Q.

It is well known that the hypotheses of Theorem 2 are invariant not only under multiplication by elliptic symbols but also under canonical change of coordinates. Also, the property under study, local solvability, is invariant under such transformations, which means that $\mathbb{P}$ is locally solvable at x_0 if and only if $U^{-1} \mathbb{P} U$ is locally solvable at $\Psi(x_0)$, where

U is an elliptic Fourier integral operator associated with the canonical transformation induced by Ψ.

As mentioned before, we shall take

$$\mathbb{P} = \Lambda^2 - c(t) D_x + d(t), \quad \Lambda = \frac{\partial}{\partial t} + b(t) D_x, \quad N = 2, \quad x_0 = (0, 0) \tag{5}$$

where b, c, d are analytic functions of t, $|t| < T$. Furthermore, we assume that $b(t)$ is *real*. The formal transpose ${}^t\mathbb{P} = \Lambda^2 + c(t) D_x + d(t)$. The hypotheses in Theorem 2 imply that

$$b(t) = b_0 t^k (1 + O(t)), \quad b_0 \text{ real} \neq 0, \quad k = 2p + 1,$$
$$p \text{ an integer} \geq 0, \tag{6}$$

and for convenience we assume that $b_0 > 0$.

We shall use the notation: $B(t) = \int_0^t b(t') \, dt'$, $z = x + iB(t)$ (note that $\Lambda z = 0$), $\tau = +\sqrt{\rho}$, and assume that $c(t)$ *is not identically zero* (otherwise the proof is even simpler).

Case I. *The lower-order terms have a strong influence*, i.e., we suppose that

$$\text{Im}\sqrt{c(t)} = c_0 t^{\ell/2}(1 + O(t)), \quad c_0 \neq 0, \quad \ell + 1 < k,$$
$$\text{say for } t > 0. \tag{7}$$

As can be observed from the proof, it is no restriction to assume that (7) holds for $t > 0$. We choose $w = w_0 + (1/\tau)w_1$, where $w_0 = z + iz^2$ (note that $\Lambda w_0 = 0$) and w_1 satisfies $(\Lambda w_1)^2 = c(t) w_{0x}$, $w_1|_{t=0} = 0$. Furthermore, we take $\varphi = g \sum_{j=0}^{J} \tau^{-j} \varphi_j$, where the φ_j are independent of ρ and satisfy in some open neighborhood of the origin

$$2(\Lambda w_1)\Lambda \varphi_j + (\Lambda^2 w_1 + c(t) D_x w_1)\varphi_j = i \, {}^t\mathbb{P}\varphi_{j-1}, \tag{8}_j$$

PSEUDODIFFERENTIAL OPERATORS   45

with the agreement that $\varphi_j \equiv 0$ if $j < 0$. As for g, it will be a C^∞ function of (x, t) whose support will be compact and contained in a suitable open set; g will, in general, depend on ρ. It will be chosen later, as will be the integer J.

Taking into account all our choices, we obtain

$$^t\mathbb{P}(e^{i\rho w}\varphi) = g(^t\mathbb{P}\varphi_J)\tau^{-J}e^{i\rho w} + L_g\left(e^{i\rho w}\sum_{j=0}^{J}\tau^{-j}\varphi_j\right) \qquad (9)$$

where $L_g = 2(\Lambda g)\Lambda + \Lambda^2 g + c(t) D_x g$ vanishes identically in an open set where $g \equiv 1$. A simple computation shows that

$$\rho \text{ Im } w = \rho(x^2 - B(t)^2) + \rho B(t) + \tau \text{ Im } w_1 \qquad (10)$$

and from the equation defining w_1, (6), and (7) we obtain

$$\text{Im } w_1 = -|c_0|\frac{t^{1+\ell/2}}{1+\ell/2}(1 + O(t)) + xt\, O(1), \qquad (11)$$

where the sign in the right-hand side of (11) was chosen so that $(\text{Im } w_1)_t < 0$ for $x = 0$ and $t > 0$. If we substitute (11) in (10) we get

$$\rho \text{ Im } w \geq \frac{1}{2}\rho x^2 + h(\rho, t) - Kt^2 \qquad (12)$$

where $K > 0$ is independent of ρ and

$$h(\rho, t) = \rho b_0 \frac{t^{k+1}}{k+1}(1 + O(t)) - \tau|c_0|\frac{t^{1+\ell/2}}{1+\ell/2}(1 + O(t)). \qquad (13)$$

Let us set $\delta = (k - \ell - 1)/(2k - \ell)$, $\varepsilon = 1/(2k - \ell)$. By (6) and (7), $0 < \varepsilon, \delta < 1$. The function h has a *minimum* for $t = t_m(\rho) = \rho^{-\varepsilon}(c_1 + o(1))$ where $c_1 = |c_0/b_0|^{2\varepsilon} > 0$ and $o(1)$ tends to zero when $\rho \to \infty$. The value of this minimum is $-\rho^\delta h_0(\rho)$ where $-h_0(\rho) = -c_2 + o(1)$, $c_2 > 0$ independent of ρ. We begin now the determination of the amplitude function φ and, first of all, of the cut-off function g. We take $g(x, t) = g_0(\rho^\varepsilon t) g_1(x)$, where g_1 is a C^∞ function on the real line, with compact support contained in a sufficiently small neighborhood of the origin and equal to one in a subneighborhood of the origin. As for $g_0(s)$, it is also a C^∞ function on the real line; but its support is contained in a small interval $(c_1 - \eta, c_1 + \eta)$ with $0 < \eta < c_1/2$, where c_1 is the positive constant entering the expression of $t_m(\rho)$. Moreover, $g_0(s)$ must be equal to *one* in the interval $(c_1 - \eta/2, c_1 + \eta/2)$. Clearly, for ρ sufficiently large, $\rho^\varepsilon t_m(\rho)$ will be an interior point of this latter interval. Next we shall study the terms φ_j in the expansion of φ by looking at the equations $(8)_j$ which they satisfy. It is convenient to regard x as a complex variable and perform the change of variables $z = x + iB(t)$, $s = t$. We have then $\partial/\partial s = \Lambda$, $\partial/\partial z = \partial/\partial x$. If we use the equation defining w_1 to rewrite $(8)_j$, we obtain, after some computation, the following inequality, valid for all pairs of integers p_0, p_1 and for x, t > 0 sufficiently near the origin

$$\left|\left(\frac{\partial}{\partial t}\right)^{p_0}\left(\frac{\partial}{\partial x}\right)^{p_1} \varphi_j(x, t)\right| \leq C_{p_0,p_1} t^{-p_0 - j(1+\ell/2) - \ell/4}. \quad (14)$$

It is also very easy to verify that

$$\left|\left(\frac{\partial}{\partial t}\right)^{p_0}\left(\frac{\partial}{\partial x}\right)^{p_1} w_1(x, t)\right| \leq C'_{p_0,p_1} t^{1-p_0}, \quad (t > 0). \quad (15)$$

We are now in a position to disprove Eq. (3). We recall that, on support of $g(x, t)$, we have $(1/2)c_1\rho^{-\varepsilon} \leq t \leq (3/2)c_1\rho^{-\varepsilon}$. Let (p_0, p_1) be any pair of nonnegative integers such that $p_0 + p_1 \leq M$. It is not difficult to see, by applying (14), (15), and (12), that, on supp g,

$$\left|\left(\frac{\partial}{\partial t}\right)^{p_0}\left(\frac{\partial}{\partial x}\right)^{p_1} (g e^{i\rho w} t_\mathbb{P} \varphi_J) \tau^{-J}\right| \leq C'_M \rho^{M+1-J\delta} \exp[\rho^\delta h_0(\rho)]. \quad (16)$$

On the other hand, again by (14), (15), and the fact that *on the support of the gradient of* g we have, by virtue of (12),

$$\rho \operatorname{Im} w \geq c_3 \rho^\delta - \rho^\delta h_0(\rho) - Kt^2, \quad c_3 > 0,$$

we obtain

$$\left|\left(\frac{\partial}{\partial t}\right)^{p_0}\left(\frac{\partial}{\partial x}\right)^{p_1} L_g\left(e^{i\rho w} \sum_{j=0}^{J} \tau^{-j}\varphi_j\right)\right| \leq C_M'' \rho^{M+1}$$
$$\times \exp\left[-c_3\rho^\delta + \rho^\delta h_0(\rho)\right]. \quad (17)$$

Clearly, for ρ sufficiently large, we will have $\exp(-c_3\rho^\delta) \leq \rho^{-J\delta}$. Also, given any integer $J' > 0$, by taking J so large as to have $J\delta \geq J' + M + 1$, we derive from (9), (16), and (17)

$$\sup_{\mathbb{R}^2} \sum_{p_0+p_1 \leq M} \left|\left(\frac{\partial}{\partial t}\right)^{p_0}\left(\frac{\partial}{\partial x}\right)^{p_1}({}^t\mathbb{P}v)\right| \leq C_M''' \rho^{-J'} \exp\left[\rho^\delta h_0(\rho)\right]. \quad (18)$$

Next we consider the left-hand side in (3). We shall choose $f(x, t) = F(\rho x, \rho(t - t_m))$, $F \in C_c^\infty(\mathbb{R}^2)$ ($\hat{F}(-\xi, 0) \neq 0$, where the caret denotes a Fourier transform). After somewhat lengthy computation we derive, for ρ sufficiently large,

$$\left|\iint fv\, dx\, dt\right| \geq \rho^{-2} \exp\left[\rho^\delta h_0(\rho)\right]. \quad (19)$$

Observing that

$$\sup_{\mathbb{R}^2} \sum_{|\alpha| \leq M} |D^\alpha f| \leq \text{const } \rho^M, \quad (20)$$

we combine (18), (19), and (20). If we choose $J' > M + 2$ and let ρ go to $+\infty$, we reach a conclusion which contradicts Eq. (3).

Case II. *The lower-order terms have little influence.*

We write $c(t) = c_0 t^r(1 + O(t))$, $c_0 \neq 0$. There are only two possible ways to negate (7), namely, either $r \geq 2p$ or else the following condition must hold:

c_0 is real > 0,

$r = 2q < 2p, \quad c(t) = c_0 t^{2q} + \cdots + c_{p-q-1} t^{p+q-1} + o(t^{p+q-1}), \quad (21)$

where c_j is real ($0 \leq j \leq p - q - 1$).

We shall assume (21) (the case $r \geq 2p$ is even simpler and is left to the reader). We shall use the same notation as before. However, we take here $w = w_0 = z + iz^2$. We get

$$e^{-i\rho w} \, {}^t\!\mathbb{P}(e^{i\rho w}\varphi) = \varphi_{tt} + \rho Z \, c(t) \, \varphi - i \, c(t) \, \varphi_z + d(t) \, \varphi, \qquad (22)$$

where $Z = 1 + 2iz$. By making a suitable (analytic) change of variables, $s = s(t)$, we many assume that

$\quad d(t)$ is purely imaginary (possibly zero) for t real. (23)

Under this assumption we are going to prove the existence and certain estimates of the solution of the Cauchy problem:

$$\varphi_{tt} + \rho Z \, c(t) \, \varphi - i \, c(t) \, \varphi_z + d(t) \, \varphi = 0, \qquad (24)$$

$$\varphi = 1, \quad \varphi_t = 0 \quad \text{at} \quad t = 0.$$

We shall adapt the proof of the Ovcyannikov theorem (see [7]) and apply it to the successive Cauchy problems:

$$\varphi_{tt}^j + \{\rho \, c(t) + d(t)\}\varphi^j = i \, c(t)\{\varphi_z^{j-1} - 2\rho z \varphi^{j-1}\}, \qquad (25)$$

$$\varphi^j = 1, \quad \varphi_t^j = 0 \quad \text{at} \quad t = 0$$

(with the agreement that $\varphi^j \equiv 0$ when $j < 0$). Let us set $C(t, \rho) = \rho c(t) + d(t)$. We shall consider Cauchy problems of the following kind:

$$u_{tt} + C(t, \rho) \, u = c(t) \, f(z, t, \rho), \qquad (26)$$

$$u = u_0, \quad u_t = 0 \quad \text{at} \quad t = 0$$

(u_0 is either 0 or 1).
 In view of (21) and (23) we have

$$C(t, \rho) = c_0 \rho t^{2q}(1 + t \, h_1(t)) + i d_0 t^{q'}(1 + t \, h_2(t)),$$

PSEUDODIFFERENTIAL OPERATORS

where h_j ($j = 1, 2$) are analytic functions near $t = 0$; h_2 is *real*, but h_1 is not necessarily so. Let us write $C(t, \rho) = C_0(t, \rho) + i\, C_1(t, \rho)$. Note that

$$C_0(t, \rho) = \rho c_0 t^{2q}(1 + O(t)), \quad C_1(t, \rho) = \rho c_1 t^{p+q}(1 + O(t)) + d_0 t^{q'}. \tag{27}$$

Let us momentarily set $g = g(t, \rho) = (1 + C_0)^{-1/2}$. If we multiply (26) by $g^2 \bar{u}_t$, take 2 Re of both members, and integrate from 0 to t, we obtain the following (note that g^2 is a decreasing function of $|t|$ in a fixed, i.e., independent of ρ, interval $|t| < T$ and that if T is sufficiently small, we have $|c| g^2 \leq 2/\rho$)

$$|gu_t|^2 + |u|^2 \leq u_0^2 + 2\int_0^t |1 - iC_1| g^2 |uu_t|\, dt' + \frac{4}{\rho}\int_0^t |fu_t|\, dt'. \tag{28}$$

We have $g \leq \text{const}\,(1 + \tau |t|^q)^{-1} \leq \text{const}\, g$, $1 + |C_1| \leq \text{const}\,(1 + \tau |t|^p)(1 + \tau |t|^q)$, whence

$$|gu_t|^2 + |u|^2 \leq u_0^2 + M_0 \int_0^t |u||gu_t|(1 + \tau |t'|^p)\, dt' + M_1 \int_0^t |f||gu_t|\, dt'\, \tau^{-1}. \tag{29}$$

<u>Estimates for φ^0</u>. If $u = \varphi^0$, we take $f \equiv 0$ and $u_0 = 1$ in the above considerations. We derive at once

$$|g\varphi_t^0|^2 + |\varphi^0|^2 \leq M_3 \exp(M_4 \tau |t|^{p+1}). \tag{30}$$

<u>Estimates for φ^j, $j > 0$</u>. In this case we take $u^j = \varphi^j - \varphi^{j-1} = u$ in (26), $u_0 = 0$, and $f(z, t, \rho) = i\{u_z^{j-1} - 2\rho z u^{j-1}\}$. We derive from (29)

$$|u| \leq w \leq M_6 \int_0^t \{|u|(1 + \tau |t'|^p) + \tau^{-1}|f|\}\, dt', \tag{31}$$

where $w(z, t, \rho) = \sup_{0 \leq t' \leq t}\{(|gu_t|^2 + |u|^2)^{1/2}(z, t', \rho)\}$.

Let us set $\Omega(t, \rho) = M_6 \int_0^t (1 + \tau|t'|^p) \, dt'$, $v = |u^j|$. Then (31) can be rewritten in short,

$$v \leq \int_0^t v\Omega' \, dt' + \int_0^t F \, dt'.$$

Let y be the solution of $y = \int_0^t y\Omega' \, dt' + \int_0^t F \, dt'$. We must have $v \leq y$ (recall that $v \geq 0$). But we know the exact expression of y:

$$y(t) = \int_0^t e^{\Omega(t)-\Omega(t')} F(t') \, dt',$$

whence

$$|v^j(z, t)| \leq 2M_6 \int_0^t \{\tau^{-1}|v_z^{j-1}(z, t')| + \tau|z||v^{j-1}(z, t')|\} \, dt', \quad (32)$$

where we have set $v^j(z, t) = e^{-\Omega(t)} u^j(z, t)$. By Cauchy's inequality,

$$|v_z^{j-1}(z, t)| \leq \frac{1}{r_j} \sup_{|z-z'|<r_j} |v^{j-1}(z', t)|. \quad (33)$$

Let $V^j(R, t) = \sup_{|z|<R}|v^j(z, t)|$. We derive from (32)

$$V^j(R, t) \leq 2M_6((\tau r_j)^{-1} + \tau R) \int_0^t V^{j-1}(R + r_j, t') \, dt'. \quad (34)$$

Let us then set $V^j(R, t) = W^j(R) (2M_6)^j t^j/j!$. We derive from (34)

$$W^j(R) \leq ((\tau r_j)^{-1} + \tau R) W^{j-1}(R + r_j) \quad (35)$$

Let us emphasize the fact that the φ^j (defined by (25)) are entire functions of z. In view of (30) we may take $W^0(R) = M$, a constant independent of R. We are then going to choose, for $j > 0$, $r_j = (\tau j)^{-1}$. We then derive from (35)

PSEUDODIFFERENTIAL OPERATORS

$$W^j(R) \leq M(j + \tau R)\left(j - 1 + \tau R + \frac{1}{j}\right)$$
$$\times \left(j - 2 + \tau R + \frac{1}{j} + \frac{1}{j-1}\right) \cdots \quad (36)$$

Let us set

$$S = \sup_j \frac{1}{j}\left(1 + \frac{1}{2} + \cdots + \frac{1}{j}\right).$$

Then

$$W^j(R) \leq Mj^j\left(1 + S + \frac{\tau R}{j}\right)^j \leq M(1 + S)^j j^j \exp((1 + S)^{-1}\tau R). \quad (37)$$

Since $j^j \leq M_7 3^j j!$, we find $V^j(R, t) \leq M_8 M_9^j |t|^j \exp(M_{10} \tau R)$. If we go back to the definition of $V^j(R, t)$ and to that of $v^j(z, t)$, we obtain

$$|u^j(z, t)| \leq M_8 (M_9 |t|)^j \exp(M_{10} \tau R + \Omega(t)),$$
$$|z| < R, \quad |t| < T. \quad (38)$$

It is important to observe that the constants M_8, M_9, M_{10} can be taken independently of R, and that R can be arbitrary. We shall take $R = |z|$. On the other hand, we have, for the solution φ of (24),

$$\varphi = \lim_{j \to +\infty} \varphi^j = \sum_{j=0}^{+\infty} u^j.$$

We derive from (38) (choosing T small enough)

$$|\varphi(z, t)| \leq \text{const} \exp(M_{10} \tau |z| + \Omega(t)). \quad (39)$$

It is not difficult to show that

$$\left|\left(\frac{\partial}{\partial t}\right)^{p_0}\left(\frac{\partial}{\partial z}\right)^{p_1} \varphi(z, t)\right| \leq C_{p_0, p_1} \rho^{p_0/2} \exp(M_{10} \tau |z| + \Omega(t)). \quad (40)$$

Indeed, for $p_0 = 0$ or 1, the combination of the method described above with the Cauchy's inequalities in the z variables gives the result. For $p_0 > 1$ it suffices to use the differential equation (24), differentiated as many times as we wish with respect to both z and t.

From here on, the argument to disprove (3) is exactly the same as in the proof of Theorem 6.1.1 in [8] and we leave it to the reader. In particular, we take $f(x, t) = \rho^2 F(\rho x, \rho t)$, where $F \in C_c^\infty(\mathbb{R}^2)$ satisfies $\iint e^{i<\xi, x>} F(x, t) \, dx \, dt = 1$.

REFERENCES

1. F. Cardoso and F. Treves, A necessary condition of local solvability for pseudodifferential equations with double characteristics, *Ann. Inst. Fourier*, 24(1974), 225-295.

2. L. Nirenberg and F. Treves, On local solvability of linear partial differential equations. Part I: Necessary conditions, *Comm. Pure Appl. Math.*, 23(1970), 1-38; Part II: Sufficient conditions, *Comm. Pure Appl. Math.*, 23(1970), 459-510.

3. J. Sjöstrand, Une classe d'opérateurs pseudodifferentiels à caractéristiques multiples, *C. R. Acad. Sci. Paris*, t. 275 (1972), 817-819.

4. L. Hörmander, Pseudodifferential operators and nonelliptic boundary problems, *Ann. Math.*, 83(1966), 129-209.

5. L. Hörmander, Pseudodifferential operators, *Comm. Pure Appl. Math.*, 18(1965), 501-517.

6. F. Cardoso and F. Treves, On subelliptic pseudodifferential operators, *Proc. Colloq. Analysis, Rio de Janeiro* (Brazil), 1972, Hermann, Paris, 1975.

7. F. Treves, Ovcyannikov theorem and hyperdifferential operators, *Notas de Matemática*, IMPA, Rio de Janeiro (Brazil), 1968.

8. L. Hörmander, *Linear Partial Differential Operators*, Springer, Berlin, 1963.

ENERGY CONCENTRATION FOR FIRST-ORDER HYPERBOLIC SYSTEMS

David G. Costa[*]

Instituto de Matemática
Universidade Federal do Rio de Janeiro
Rio de Janeiro, Brazil

1. INTRODUCTION

In this chapter we consider first-order hyperbolic systems of k equations in $\mathbb{R}^n$ (n odd ≥ 3),

$$\frac{\partial u}{\partial t} + \sum_{j=1}^{n} A_j \frac{\partial u}{\partial x_j} = 0; \qquad (1.1)$$

i.e., $A = (A_1, \ldots, A_n)$ is an n-tuple of $k \times k$ matrices with constant complex coefficients such that the eigenvalues $\lambda_1(\xi), \ldots, \lambda_k(\xi)$ of $A \cdot \xi = \sum_{j=1}^{n} \xi_j A_j$ are real for all $\xi = (\xi_1, \ldots, \xi_n) \in \mathbb{R}^n \setminus \{0\}$.

---

[*] Present address: Departmento de Matemática, Universidade de Brasília, Brasília-D. F., Brazil.

We aim to study the behavior of solutions along "nonbicharacteristic" rays.* We show that solutions subject to "nonstationary" data decay along such rays as $O(t^{-n})$ as $t \to \infty$. We also consider the decay of energy in nonbicharacteristic cones and obtain more precise results in the isotropic case. As we shall see, energy tends to "concentrate" around the "wave" (bicharacteristic) cone.

These results were obtained in joint work with C. Bardos (see [1]). Other results on energy decay can be found in [2] and [3], where P. D. Lax and R. S. Phillips show the decay of energy in bounded sets of $\mathbb{R}^n$. Recently C. H. Wilcox has proved results for the d'Alembert wave equation which are related to ours ([4]).

Our method uses as a primary tool the Radon transform, that is, the plane-wave decomposition of solutions ([5]). This approach proves to be very useful in having a closer look at the "Riemann" matrix of (1.1).

We shall be concerned with smoothly diagonalizable systems. In other words, there exists a smooth (C^∞) matrix $P(\omega)$, $\omega \in S^{n-1}$, such that

$$(H_1) \qquad P(\omega) \, A \cdot \omega P(\omega)^{-1} = \begin{pmatrix} \lambda_1(\omega) & & \bigcirc \\ & \ddots & \\ \bigcirc & & \lambda_k(\omega) \end{pmatrix}$$

$$(H_2) \qquad \dim(\ker A \cdot \omega) = \text{const} \quad \text{for all} \quad \omega \in S^{n-1}.$$

In the following section we define and review a few facts about the Radon transform and use them to derive a solution formula for the Cauchy problem for (1.1).

---

* Here we reproduce part of a series of lectures on decay for hyperbolic systems delivered at Tulane University during the period January-February 1974.

2. THE RADON TRANSFORM AND THE CAUCHY PROBLEM

Given a function $f \in S(\mathbb{R}^n)$ (the space of rapidly decreasing functions), the Radon transform of f, $\hat{f}$ (or Rf), is defined by the formula

$$\hat{f}(s, \omega) = \int_{x \cdot \omega = s} f(x) \, dx, \quad s \in \mathbb{R}, \quad \omega \in S^{n-1}.$$

One can show that $\hat{f} \in S(\mathbb{R} \times S^{n-1})$ (see [6, 7]). The following properties are immediate:

(i) $\hat{f}$ is an even function on $\mathbb{R} \times S^{n-1}$.

(ii) For each integer $\ell \geq 0$, $\int_{-\infty}^{\infty} s^\ell \hat{f}(s, \omega) \, ds$ can be written as a polynomial in ω of degree ℓ.

(iii) $\left(\frac{\partial}{\partial x_j} f \right)^{\wedge} = \omega_j \frac{\partial}{\partial s} \hat{f}$.

In fact, R establishes a homeomorphism of $S(\mathbb{R}^n)$ onto the subspace of $S(\mathbb{R} \times S^{n-1})$ consisting of all functions $\varphi(s, \omega)$ satisfying (i) and (ii) above, and we have the inverse formula (see [6, 7]):

$$f(x) = c_n \int_{S^{n-1}} \left(\frac{\partial}{\partial s} \right)^{n-1} \hat{f}(x \cdot \omega, \omega) \, d\omega \quad (n \text{ odd} \geq 3)$$

$$= c_n \Delta^{(n-1)/2} \int_{S^{n-1}} \hat{f}(x \cdot \omega, \omega) \, d\omega, \quad (2.1)$$

where c_n is a constant depending only on n.

Now let $u(\cdot, t)$ be the solution of the Cauchy problem for (1.1) with initial datum $f \in C_0^\infty(\mathbb{R}^n, \mathbb{C}^k)$. Then, in view of property (iii), the Radon transform of $u(\cdot, t)$, $\hat{u}(s, \omega; t)$, is the solution of the one-dimensional Cauchy problem (ω fixed)

$$\frac{\partial \hat{u}}{\partial t} + A \cdot \omega \frac{\partial \hat{u}}{\partial s} = 0, \quad \hat{u}(s, \omega; 0) = \hat{f}(s, \omega).$$

Putting $\varphi(s, \omega; t) = P(\omega) \hat{u}(s, \omega; t)$, $P(\omega) = [p_{ij}(\omega)]$,

$P(\omega)^{-1} = [q_{ij}(\omega)]$ [see (H_1)], we obtain, for each $\ell = 1, \ldots, k$,

$$\frac{\partial \varphi_\ell}{\partial t} + \lambda_\ell(\omega) \frac{\partial \varphi_\ell}{\partial s} = 0, \quad \varphi_\ell(s, \omega; 0) = \sum_{j=1}^{k} p_{\ell j}(\omega) \hat{f}_j(s, \omega),$$

whose solution is clearly given by

$$\varphi_\ell(s, \omega; t) = \sum_{j=1}^{k} p_{\ell j}(\omega) \hat{f}_j(s - \lambda_\ell(\omega)t, \omega).$$

This yields

$$\hat{u}_i(s, \omega; t) = \sum_{\ell, j=1}^{k} q_{i\ell}(\omega) p_{\ell j}(\omega) \hat{f}_j(s - \lambda_\ell(\omega)t, \omega),$$

$$i = 1, \ldots, k,$$

and hence,

$$\hat{u}(s, \omega; t) = \sum_{\ell=1}^{k} G_\ell(\omega) \hat{f}(s - \lambda_\ell(\omega)t, \omega), \qquad (2.2)$$

where $G_\ell(\omega) = [g_{ij}^{\ell}(\omega)] = [q_{i\ell}(\omega) p_{\ell j}(\omega)]$.

Formula (2.2) relates the Radon transform of $u(\cdot, t)$ and that of the initial datum f. In order to obtain $u(x, t)$ explicitly we use (2.1), which yields

$$u(x, t) = c_n \Delta^{(n-1)/2} \sum_{\ell=1}^{k} \int_{S^{n-1}} G_\ell(\omega) \left(\int_{y \cdot \omega = x \cdot \omega - \lambda_\ell(\omega)t} f(y) \, dy \right) d\omega.$$

$$(2.3)$$

3. DECAY ALONG NONBICHARACTERISTIC RAYS; ENERGY DECAY

For each $\omega \in S^{n-1}$, $\ell = 1, \ldots, k$, let $\delta(\omega, \lambda_\ell(\omega), \cdot)$ denote the temperate distribution

ENERGY CONCENTRATION

$$f \longmapsto \int_{y \cdot \omega = \lambda_\ell(\omega)} f(y) \, dy$$

and put

$$\langle I_\ell, f \rangle = \left\langle \int_{S^{n-1}} G_\ell(\omega) \, \delta(\omega, \lambda_\ell(\omega), \cdot) \, d\omega, f \right\rangle.$$

Then it is not hard to obtain from (2.3) the following:

LEMMA 3.1.

$$u(\cdot, t) = c_n \Delta^{(n-1)/2} \left[\sum_{\lambda_\ell = 0} I_\ell * f + \frac{1}{t} \sum_{\lambda_\ell \neq 0} I_\ell(\frac{\cdot}{t}) * f \right], \quad t > 0.$$

Here $I_\ell(\cdot/t)$ denotes "dilation by $1/t$ of the distribution I_ℓ" which is defined, as usual, by $\langle I_\ell(\cdot/t), g \rangle = t^n \langle I_\ell, g(t \cdot) \rangle$.

We also remark that hypothesis (H_2) in Sec. 1 implies that, for each $\ell = 1, \ldots, k$, $\lambda_\ell(\omega)$ is either identically zero or else never vanishes on S^{n-1}, a fact that justifies our summation notation above.

So, for each $\lambda_\ell \neq 0$, let us denote by $\partial \Lambda_\ell$ the smooth surface $\{\lambda_\ell(\omega)^{-1} \omega \mid \omega \in S^{n-1}\}$, by Λ_ℓ the closure of the bounded component of $\mathbb{R}^n \setminus \partial \Lambda_\ell$, and by ∂H_ℓ the polar of $\partial \Lambda_\ell$, i.e., the hull of the hyperplanes $\{x \mid x \cdot \omega = \lambda_\ell(\omega)\}$, $\omega \in S^{n-1}$ (see [8]). From now on we shall assume

(H_3) $\partial \Lambda_\ell$ is strictly convex for each $\lambda_\ell \neq 0$.

Then we obtain the following regularity result on the distributions I_ℓ corresponding to $\lambda_\ell \neq 0$ (see [9] and the Appendix).

PROPOSITION 3.2. I_ℓ is bounded together with all of its derivatives up to order $(n-3)/2$, i.e., $I_\ell \in W^{(n-3)/2, \infty}(\mathbb{R}^n, \mathbb{C}^{k^2})$; in fact, we have $I_\ell \equiv 0$ in $H_\ell = \{\xi \mid S_\ell(\xi) \leq 1\}$ and the estimates

$$|D^\alpha I_\ell(y)| \leq \frac{C_\alpha}{|y|^{1+|\alpha|}} \left(\frac{S_\ell(y) - 1}{|y|}\right)^{[(n-3)/2]-|\alpha|}$$

hold for any multi-integer α and any $y \notin H_\ell$, where $S_\ell(\xi) = \sup_{x \in \Lambda_\ell} \xi \cdot x$ is the support function of Λ_ℓ.

Now, in view of this result, we can rewrite $u(x, t)$ in Lemma 3.1 as

$$u(x, t) = c_n \Delta^{(n-1)/2} \left[\sum_{\lambda_\ell = 0} (I_\ell * f)(x) + \frac{1}{t} \sum_{\lambda_\ell \neq 0} \int_{\mathbb{R}^n} I_\ell\left(\frac{y}{t}\right) \times f(x - y) \, dy \right]. \qquad (3.1)$$

Since the first term in the bracket above does not depend on t, we can only expect decay for solutions whose initial data satisfy $\sum_{\lambda_\ell = 0} I_\ell * f = 0$. Such f's are said to be "nonstationary" for the operator $A \cdot \nabla = \sum_{j=1}^n A_j \, \partial/\partial x_j$. Nonstationary data (and only these) give rise to solutions that decay in L^r norm (r big $\leq \infty$) as $t \to +\infty$. In fact, we have proved the following result in [9].

THEOREM 3.3. Assume $(H_1)-(H_3)$. Then, the solution $u(\cdot, t)$ of (1.1) with nonstationary datum $f \in C_0^\infty(\mathbb{R}^n, \mathbb{C}^k)$ satisfies

$$\|u(\cdot, t)\|_r \leq C_q t^{-(n-1)/2 + n/q} \|f\|_{(n+1)/2, p} \quad \forall \, t > 0, \qquad (3.2)^\dagger$$

where $1/r = 1/p + 1/q - 1 \geq 0$, $\infty \geq q > 2n/(n-1)$, and $C_q > 0$ does not depend on f. In particular, we have the uniform decay

$$\|u(\cdot, t)\|_\infty \leq C_\infty t^{-(n-1)/2} \|f\|_{(n+1)/2, 1} \quad \forall \, t > 0. \qquad (3.3)$$

[†] We denote by $\|\cdot\|_{m,p}$ the usual norm in the Sobolev space $W^{m,p}(\mathbb{R}^n, \mathbb{C}^k)$; $\|\cdot\|_r = \|\cdot\|_{0,r}$ denotes the norm in $L^r(\mathbb{R}^n, \mathbb{C}^k)$.

ENERGY CONCENTRATION

Let us now consider a "nonbicharacteristic" ray $\{(t\xi, t) \mid t > 0\}$, i.e., such that $\xi \notin \partial H = \bigcup_{\lambda_\ell \neq 0} \partial H_\ell$. Denoting by $d = d(\xi, \partial H)$ the distance from ξ to the set ∂H, we obtain

THEOREM 3.4. If $f \in L^1(\mathbb{R}^n, \mathbb{C}^k)$ is nonstationary[†] with support in the ball of radius R, then

$$|u(t\xi, t)| \leq Ct^{-n}\|f\|_1 \qquad (3.4)$$

for all $t > 2R/d$.

Proof. We first prove (3.4) assuming that $f \in C_0^\infty(\mathbb{R}^n, \mathbb{C}^k)$. The result is then obtained by passing to the limit.

We have:

$$\int_{\mathbb{R}^n} I_\ell\left(\frac{y}{t}\right) f(x - y) \, dy = t^n \int_{\mathbb{R}^n} I_\ell(z) f(x - tz) \, dz.$$

Let $0 < \eta < (d/2) - (R/t)$. For $|x - t\xi| < t\eta$, the support of $f(x - t\cdot)$ is contained in the ball of center ξ and radius $d/2$. But in this ball I_ℓ is smooth, and therefore we have

$$\left|\Delta^{(n-1)/2} \int_{\mathbb{R}^n} I_\ell\left(\frac{y}{t}\right) f(x - y) \, dy\right|$$

$$= \left|t^n \int_{\mathbb{R}^n} I_\ell(z) \Delta_x^{(n-1)/2} f(x - tz) \, dz\right|$$

$$\leq t^n t^{-(n-1)} \int_{|z-\xi|\leq d/2} |\Delta_z^{(n-1)/2} I_\ell(z)| |f(x - tz)| \, dz$$

$$\leq c(\ell) \, t^{-(n-1)} t^n \int_{\mathbb{R}^n} |f(x - tz)| \, dz,$$

[†]For a distribution $f \in S'(\mathbb{R}^n, \mathbb{C}^k)$, we say that f is nonstationary iff $\langle f, \varphi \rangle = 0$ whenever $\varphi \in S(\mathbb{R}^n, \mathbb{C}^k)$ is such that $A^* \cdot \nabla \varphi = 0$. It is easy to see that the definition given previously is equivalent to this one.

provided $|x - t\xi| < t\eta$. In particular, for $x = t\xi$ and in view of (3.1) we obtain

$$|u(t\xi, t)| \leq C t^{-n} \int_{\mathbb{R}^n} t^n |f(t\xi - tz)| \, dz = Ct^{-n} \|f\|_1.$$

The proof of Theorem 3.4 is complete.

This result allows us to derive the following estimate on the energy decay in any "nonbicharacteristic" cone, i.e., any cone $\bigcup_{t>0} t\Sigma = \{(t\sigma, t) \mid \sigma \in \Sigma, t > 0\}$ over a set Σ which does not intersect ∂H.

COROLLARY 3.5. Let $f \in L^1(\mathbb{R}^n, \mathbb{C}^k)$ be nonstationary with support in the ball of radius R. Let $\Sigma \subset \mathbb{R}^n \setminus \partial H$ be a closed measurable set and $d = d(\Sigma, \partial H)$ the distance between Σ and ∂H. Then, for all $t > 2R/d$, we have

$$\int_{t\Sigma} |u(x, t)|^2 \, dx \leq Ct^{-n} \|f\|_1^2.$$

where C is a constant depending only on Σ and $A \cdot \nabla$.

<u>Proof</u>. By the finite speed of propagation property, the support of $u(\cdot, t)$ is contained in the ball of radius $R + \lambda_{max} t$, where

$$\lambda_{max} = \max_{\substack{1 \leq \ell \leq k \\ \omega \in S^{n-1}}} |\lambda_\ell(\omega)|.$$

Therefore, for $t > 2R/d$,

$$\int_{t\Sigma} |u(x, t)|^2 \, dx = \int_{t\Sigma \cap B(0, R+\lambda_{max} t)} |u(x, t)|^2 \, dx$$

$$\leq C_1 t^n \sup_y |u(ty, t)|^2,$$

ENERGY CONCENTRATION

where $C_1 = \int_{B(0,\lambda_{max}+d/2)} 1 \, dy$ and the supremum is taken over $\Sigma \cap B(0, \lambda_{max} + d/2)$.

We now cover $\Sigma \cap B(0, \lambda_{max} + d/2)$ with balls of radius $d/4$ and use Theorem 3.4 to obtain the result.

4. THE HERMITIAN, ISOTROPIC CASE

A sharper version of the above result on energy decay is obtained when we restrict ourselves to the Hermitian, isotropic (spherically symmetric) case. So, we shall assume now that the matrices A_j, $j = 1, \ldots, n$, are Hermitian. Then the principle of conservation of energy holds, namely, if $f \in L^2(\mathbb{R}^n, \mathbb{C}^k)$,

$$\int_{\mathbb{R}^n} |u(x, t)|^2 \, dx = \int_{\mathbb{R}^n} |f(x)|^2 \, dx \quad \text{for all } t.$$

The assumption of spherical symmetry means that the eigenvalues $\lambda_\ell(\omega)$ are independent of $\omega \in S^{n-1}$. Since $A \cdot \omega$ is an odd function of ω, for any λ_ℓ, $-\lambda_\ell$ is also an eigenvalue. Therefore we will write $\mu_j = |\lambda_\ell|$ and have the inequality

$$0 \leq \mu_1 < \mu_2 < \cdots < \mu_r,$$

where r ($\leq k/2$) depends on the multiplicity of each eigenvalue.

THEOREM 4.1. Let $0 < \mu_j < \mu_{j+1}$ be two successive strictly positive absolute values of eigenvalues of $A \cdot \omega$. Let $f \in L^1(\mathbb{R}^n, \mathbb{C}^k)$ be nonstationary with support in the ball of radius R. Then, if $\rho > 2R$, $\mu_j T > 2R$, $T > (\rho + R)/(\mu_{j+1} - \mu_j)$, we have

$$\int_{\Sigma_j(T,\rho)} |u(x, T)|^2 \, dx \leq \frac{C}{\rho^n} \|f\|_1^2 \qquad (4.1)$$

where $\Sigma_j(T, \rho) = \{x \mid \rho + \mu_j T < |x| < \mu_{j+1} T - R\}$.

For the proof of this theorem, we shall use the following lemma (see the proof in the Appendix) on the representation of the distribution I_ℓ, when $\lambda_\ell = 0$ (see Sec. 3 for notations).

LEMMA 4.2. The distribution $h \mapsto \int_{S^{n-1}} G_\ell(\omega) \hat{h}(0, \omega) \, d\omega$ is given by the function $(1/|x|) \tilde{G}_\ell(x/|x|)$, where $\tilde{G}_\ell$ is defined by
$$\tilde{G}_\ell(\omega) = \int_{S^{n-1} \cap \{\sigma \mid \sigma \cdot \omega = 0\}} G_\ell(\sigma) \, d\sigma.$$

Proof of Theorem 4.1. We have
$$u(x, t) = c_n \Delta^{(n-1)/2} \left[\sum_{\lambda_\ell \neq 0} \int_{S^{n-1}} G_\ell(\omega) \hat{f}(x \cdot \omega - \lambda_\ell t, \omega) \, d\omega \right].$$

Since the support of f is contained in the ball of radius R, the support of $\hat{f}$ is contained in the set $[-R, R] \times S^{n-1}$. On the other hand, for $|\lambda_\ell| \geq \mu_{j+1}$ and $x \in \Sigma_j(t, \rho)$ we have
$$|x \cdot \omega - \lambda_\ell t| \geq |\lambda_\ell| t - |x| > R.$$

Therefore, for $x \in \Sigma_j(t, \rho)$, we obtain
$$u(x, t) = c_n \sum_{|\lambda_\ell| \leq \mu_j} \Delta^{(n-1)/2} \int_{S^{n-1}} G_\ell(\omega) \hat{f}(x \cdot \omega - \lambda_\ell t, \omega) \, d\omega$$
$$= c_n \sum_{|\lambda_\ell| \leq \mu_j} u_\ell(x, t). \tag{4.2}$$

Henceforth, we will denote by λ_ℓ any eigenvalue such that $0 < |\lambda_\ell| \leq \mu_j$ and look at the corresponding u_ℓ [see (4.2)].

Using the relation $(\Delta f)\hat{\,} = (\partial^2/\partial s^2)\hat{f}$ and differentiating u_ℓ twice with respect to t, we see that u_ℓ is a solution of the wave equation $(\partial^2 u_\ell/\partial t^2) - \lambda_\ell^2 \Delta u_\ell = 0$. Therefore, if we put
$$v_\ell(x) = -\frac{1}{\lambda_\ell} \int_{S^{n-1}} \Delta^{[(n-1)/2]-1} G_\ell(\omega) \frac{\partial \hat{f}}{\partial s}(x \cdot \omega, \omega) \, d\omega$$

and define

ENERGY CONCENTRATION

$$\omega_\ell(x, t) = \int_0^t u_\ell(x, \tau) \, d\tau + v_\ell(x),$$

$\omega_\ell(x, t)$ is the solution of the wave equation $(\partial^2 \omega_\ell / \partial t^2) - \lambda_\ell^2 \Delta \omega_\ell = 0$ in $\mathbb{R}^n \times \mathbb{R}_t$, with initial data $\omega_\ell(x, 0) = v_\ell(x)$, $(\partial \omega_\ell / \partial t)(x, 0) = u_\ell(x, 0)$.

Now, by Lemma 4.2, we have

$$u_\ell(x, 0) = \Delta^{(n-1)/2} \int_{\mathbb{R}^n} \tilde{G}_\ell\left(\frac{y}{|y|}\right) \cdot \frac{1}{|y|} f(x - y) \, dy.$$

Since the function $y \mapsto \tilde{G}_\ell(y/|y|)$ is smooth and homogeneous of degree zero, it is not hard to see that, for every multi-integer α, we have the estimate

$$\left| D^\alpha \tilde{G}_\ell\left(\frac{y}{|y|}\right) \right| \leq C_\alpha |y|^{-|\alpha|}, \quad y \neq 0,$$

where C_α is a constant independent of y. Therefore, using Leibniz's formula, we obtain

$$\left| \Delta^{(n-1)/2} \frac{1}{|y|} \tilde{G}_\ell\left(\frac{y}{|y|}\right) \right| \leq C'_\ell |y|^{-n},$$

so that, for $|x| > 2R$,

$$|u_\ell(x, 0)| \leq C'_\ell \int_{|y| > |x|/2} |y|^{-n} |f(x - y)| \, dy$$

$$\leq C_\ell |x|^{-n} \|f\|_1. \tag{4.3}$$

Similarly we obtain the following estimate on ∇v_ℓ for $|x| > 2R$:

$$|\nabla v_\ell(x)| \leq D_\ell |x|^{-n} \|f\|_1. \tag{4.4}^\dagger$$

---

† The constants C_ℓ and D_ℓ appearing in (4.3) and (4.4) are independent of f and x.

Hence the function $\omega_\ell(x, t)$ is of finite energy and we obtain the following, using the finite speed of propagation property and recalling that $|\lambda_\ell| \leq \mu_j$:

$$\int_{|x|\geq\rho+\mu_j T} |u_\ell(x, T)|^2 \, dx = \int_{|x|\geq\rho+\mu_j T} \left|\frac{\partial\omega_\ell}{\partial t}(x, T)\right|^2 \, dx$$

$$\leq \int_{|x|\geq\rho+|\lambda_\ell|T} \left(\left|\frac{\partial\omega_\ell}{\partial t}(x, T)\right|^2 + \lambda_\ell^2 |\nabla\omega_\ell(x, T)|^2\right) dx$$

$$\leq \int_{|x|\geq\rho} (|u_\ell(x, 0)|^2 + \lambda_\ell^2 |\nabla v_\ell(x)|^2) \, dx$$

$$\leq C\rho^{-n} \|f\|_1^2$$

[by (4.3) and (4.4)]. The proof is complete.

Theorem 4.1 shows that the energy of the solution "concentrates along characteristics" (in the isotropic case the notions of characteristic and bicharacteristic coincide): namely, that for ρ large enough the energy in the expanding shells $\Sigma_j(T) = S(\mu_j, \mu_{j+1}, \rho) \cap \{(x, T)\}$,[†] $T \nearrow \infty$, is arbitrarily small (see Fig. 1).

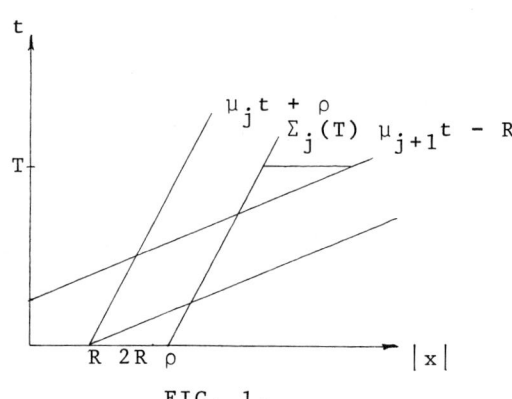

FIG. 1.

[†] $S(\mu_j, \mu_{j+1}, \rho)$ is defined as $\{(x, t) \mid t > (\rho + R) \div (\mu_{j+1} - \mu_j), \mu_j t + \rho \leq |x| \leq \mu_{j+1} t - R\}$.

Combining this result with the conservation of energy we obtain the following result, which shows the optimality of the uniform decay rate given by Theorem 3.3 [see (3.3)]:

THEOREM 4.3. Assume that the matrices A_i are Hermitian and that the eigenvalues of $A \cdot \omega$ are independent of ω. Let $f \in C_0^\infty(\mathbb{R}^n, \mathbb{C}^k)$ be a nonstationary function (not identically zero). Then there exists a constant $C > 0$ (independent of t) such that

$$\sup_{x \in \mathbb{R}^n} |u(x, t)| \geq C t^{-(n-1)/2} \quad \text{for } t \text{ large}.$$

Proof. Choose ρ as in Theorem 4.1 and such that the right-hand side of (4.1) is $\leq E/2k$, where $E = \|u(\cdot, t)\|_2^2 = \|f\|_2^2$ is the total energy. Then, for t large,

$$E = \sum_j \int_{\Sigma_j(t,\rho)} |u(x, t)|^2 \, dx + E_1(t) \leq \frac{E}{2} + E_1(t), \qquad (4.5)$$

where $E_1(t)$ denotes the energy in the remaining spherical shells (around the characteristics). But, as these latter have constant thickness (independent of t), we have $E_1(t) \leq C_1 \sup_{x \in \mathbb{R}^n} |u(x,t)|^2 \times t^{n-1}$ for some constant $C_1 > 0$. The result follows in view of (4.5).

5. APPENDIX

Here we prove Proposition 3.2 in the isotropic case and Lemma 4.2, which describe the distributions I_ℓ corresponding to $\lambda_\ell \neq 0$ and $\lambda_\ell = 0$, respectively.

Proof of Proposition 3.2. We are assuming $\lambda_\ell(\omega) = \lambda_\ell$ $\forall \, \omega \in S^{n-1}$. Recalling the definitions of Sec. 3, we have

$$\partial \Lambda_\ell = \{\lambda_\ell^{-1}\omega \mid \omega \in S^{n-1}\},$$

$$S_\ell(\xi) = |\lambda_\ell|^{-1}|\xi|,$$

$$H_\ell = \{\xi \mid |\xi| \leq |\lambda_\ell|\}$$

and

$$\langle I_\ell, f \rangle = \int_{S^{n-1}} G_\ell(\omega) \langle \delta(\omega, \lambda_\ell, \cdot), f \rangle \, d\omega$$

$$= \int_{S^{n-1}} G_\ell(\omega) \, \hat{f}(\lambda_\ell, \omega) \, d\omega.$$

We now let

$$H(\sigma) = 0, \quad \sigma < 0$$
$$= 1, \quad \sigma \geq 0$$

denote the Heaviside function and define, for each $t > 0$, the function

$$J_\ell(y, t) = \frac{d}{dt} \int_{S^{n-1}} G_\ell(\omega) \, H(t - y \cdot \omega) \, d\omega. \qquad (5.1)$$

Then, using Fubini's theorem and observing that

$$\hat{f}(t, \omega) = \frac{d}{dt} \int_{y \cdot \omega \leq t} f(y) \, dy,$$

we get

$$\int_{S^{n-1}} G_\ell(\omega) \, \hat{f}(\lambda_\ell, \omega) \, d\omega = \int_{\mathbb{R}^n} f(y) \, J_\ell(y, \lambda_\ell) \, dy,$$

i.e., the distribution I_ℓ is given by the function $J_\ell(\cdot, \lambda_\ell)$ [see (5.1)].

Clearly (5.1) gives $J_\ell(y, \lambda_\ell) = 0$ for $|y| \leq |\lambda_\ell|$. Now we estimate $J_\ell(y, \lambda_\ell)$ and its derivatives for $|y| > |\lambda_\ell|$. Without loss of generality we may put $G_\ell(\omega) =$ identity in (5.1), so that writing $\int_{S^{n-1}} H(t - y \cdot \omega) \, d\omega$ as an iterated integral yields

$$\int_{S^{n-1}} H(t - y \cdot \omega) \, d\omega = \Omega_{n-2} \int_{-1}^{1} H(t - \tau|y|)(1 - \tau^2)^{(n-3)/2} \, d\tau$$

$$= \Omega_{n-2} \int_{-1}^{t/|y|} (1 - \tau^2)^{(n-3)/2} \, d\tau$$

for $|y| > t$ (where Ω_{n-2} = surface measure of S^{n-2}), and hence

$$I_\ell(y) = J_\ell(y, \lambda_\ell) = \Omega_{n-2} \frac{1}{|y|} \left(1 - \frac{\lambda_\ell^2}{|y|^2}\right)^{(n-3)/2}$$

for $|y| > |\lambda_\ell|$.

Finally, using Leibniz's formula, we obtain the estimates

$$|D^\alpha I_\ell(y)| \leq \frac{C_\alpha}{|y|^{1+|\alpha|}} \left(1 - \frac{|\lambda_\ell|}{|y|}\right)^{[(n-3)/2]-|\alpha|},$$

for $|y| > |\lambda_\ell|$, any α,

which show (together with $I_\ell(y) = 0$, $|y| < |\lambda_\ell|$), in particular, that $I_\ell \in W^{(n-3)/2,\infty}$.

Proof of Lemma 4.2. We wish to show that, for $\lambda_\ell = 0$, the distribution $I_\ell: h \mapsto \int_{S^{n-1}} G_\ell(\omega) \, \hat{h}(0, \omega) \, d\omega$ is given by the function

$$\frac{1}{|x|} \tilde{G}_\ell\left(\frac{x}{|x|}\right),$$

where

$$\tilde{G}_\ell(\omega) = \int_{S^{n-1} \cap \{\sigma | \sigma \cdot \omega = 0\}} G_\ell(\sigma) \, d\sigma, \quad \omega \in S^{n-1}.$$

Let $\omega_0 \in S^{n-1}$ be fixed and denote by $R(\omega)$ the rotation leaving the "axis" $\{y \mid y \cdot \omega_0 = 0\} \cap \{y \mid y \cdot \omega = 0\}$ invariant, such that $R(\omega) \omega_0 = \omega$. Then, for any $h \in S(\mathbb{R}^n, \mathbb{C}^k)$,

$$\langle I_\ell, h \rangle = \int_{S^{n-1}} G_\ell(\omega) \left(\int_{y \cdot \omega = 0} h(y) \, dy\right) d\omega$$

$$= \int_{S^{n-1}} G_\ell(R(\omega)\,\omega_0) \left[\int_{\mathbb{R}^{n-1}} h(R(\omega)\,y)\,dy \right] d\omega$$

$$= \int_{\mathbb{R}^{n-1}} \left[\int_{S^{n-1}} G_\ell(R(\omega)\,\omega_0)\,h(R(\omega)\,y)\,d\omega \right] dy,$$

by Fubini's theorem. Using polar coordinates ($y = r\theta$), we get

$$\langle I_\ell, h \rangle = \int_0^\infty r^{n-2} \int_{S^{n-2}} \left[\int_{S^{n-1}} G_\ell(R(\omega)\,\omega_0) \right.$$
$$\left. \times h(r\,R(\omega)\,\theta)\,d\omega \right] d\theta\,dr$$

$$= \int_0^\infty r^{n-2} \int_{S^{n-2}} \left[\int_{S^{n-1}} \overline{G}_\ell(\overline{\omega},\theta)\,h(r\overline{\omega})\,d\overline{\omega} \right] d\theta\,dr,$$

where $\overline{\omega} = R(\omega)\,\theta$ and $\overline{G}_\ell(\overline{\omega},\theta) = G_\ell(R(\omega)\,\omega_0)$. Now

$$\int_{S^{n-2}} \overline{G}_\ell(\overline{\omega},\theta)\,d\theta = \widetilde{G}_\ell(\overline{\omega}),$$

so that using Fubini's theorem once more, we finally obtain

$$\langle I_\ell, h \rangle = \int_0^\infty r^{n-2} \left[\int_{S^{n-1}} \widetilde{G}_\ell(\omega)\,h(r\omega)\,d\omega \right] dr$$

$$= \int_{\mathbb{R}^n} \frac{1}{|x|} \widetilde{G}_\ell\!\left(\frac{x}{|x|}\right) h(x)\,dx.$$

The proof is complete.

REFERENCES

1. C. Bardos and D. G. Costa, Décroissance de la solution d'un Système Hyperbolique du Premier Order Diagonalisable, *C. R. Acad. Sci. Paris*, 278 (1974), 265-267.

2. P. D. Lax and R. S. Phillips, Scattering theory, *Bull. Amer. Math. Soc.*, 70 (1964), 130-142.

3. P. D. Lax and R. S. Phillips, *Scattering Theory*, Academic Press, New York, 1967.

4. C. H. Wilcox, *Asymptotic Wave Functions and Energy Distributions for the d'Alembert Wave Equation*, Univ. of Utah, Salt Lake City, 1974.

5. F. John, *Plane Waves and Spherical Means Applied to Partial Differential Equations*, Wiley (Interscience), New York, 1955.

6. S. Helgason, The Radon transform on Euclidean spaces ..., *Acta Math.*, *113* (1965), 93-106.

7. D. Ludwig, The Radon transform on Euclidean spaces, *Comm. Pure Appl. Math.*, *19* (1966), 49-81.

8. H. Buseman, *Convex Surfaces*, Wiley (Interscience), New York, 1958.

9. D. G. Costa, *The Uniform Behavior of Solutions of Linear Hyperbolic Equations for Large Times*, Ph. D. Dissertation, Brown Univ., Rhode Island, 1973.

EXISTENCE THEOREMS IN HIGHER DIMENSIONAL
CALCULUS OF VARIATIONS*

Ubiratan D'Ambrosio

Instituto de Matemática
Universidade Estadual de Campinas
Campinas, São Paulo, Brazil

Since the 1950's, existence and regularity for multiple integral problems of the Calculus of Variations has experienced enormous progress due to the introduction of the so-called geometric measure theoretic methods.

In this chapter we will first survey the background of these methods and then indicate some general existence theorems.

1. RECTIFIABLE SETS

The basic concept is that of a rectifiable set. We say that $A \subset \mathbb{R}^n$ is a countably m-rectifiable set if A is expressible

*Research partially supported by Grant FAPESP 73/740.

as $\bigcup_{j=1}^{\infty} A_j$, where for each j, A_j is contained in an m-dimensiomal submanifold of $\mathbb{R}^n$ of class 1. Rectifiable sets were discussed in [1].

Two measures frequently encountered are the m-dimensional Hausdorff measure $H^m(S)$ and the m-dimensional integral geometric measure $I^m(S)$. They are defined as follows:

The Hausdorff m-dimensional measure of a set S of $\mathbb{R}^n$ is defined as

$$H^m(S) = \lim_{r \to 0^+} \inf \left\{ \sum_{i=1}^{\infty} g_m(S_i) : \bigcup_{i=1}^{\infty} S \supset S; \text{ diam } S_i < r, i = 1, 2, \ldots \right\}$$

where the gauge $g_m(S)$ is the product of Lebesgue measure of the m-dimensional ball of $\mathbb{R}^n$, of unitary diameter, by the diameter of S to the power m.

Let H be the space of all n - m planes in $\mathbb{R}^n$. Then the m-dimensional integral geometric measure of a set S of $\mathbb{R}^n$ is defined as

$$I^m(S) = k \int_H H^0(S \cap E) \, d\psi(E),$$

where ψ is an invariant measure and k is a constant. We remark that $H^m(S) \geq I^m(S)$.

We may now define the *Hausdorff area* of a Lipschitzian map $f: \mathbb{R}^m \to \mathbb{R}^n$ over an L^m-measurable set $A \subset \mathbb{R}^m$ as

$$\int_{\mathbb{R}^n} H^0(A \cap f^{-1}\{y\}) \, dH^m y.$$

It turns out that this area coincides with the classical area

$$\int_A J_m f \, dL^m x$$

where the Jacobian J_m is obtained in the following way: for $x \in \mathbb{R}^m$, consider $Df(x)$ and the m-vectors $\Lambda_m Df(x)$. Then $J_m f(x) = \|\Lambda_m Df(x)\|$, where $\|g\| = \sup\{|g(x)|: x \in V \text{ and } |x| \leq 1\}$ for any linear map $g: V \to V'$, V and V' finite dimensional vector spaces.

CALCULUS OF VARIATIONS

These results can be generalized by introducing the concept of *slicing*. Put

$$Y_i = \{y: H^{m-i}(A \cap f^{-1}\{y\}) > 0\}$$

and

$$X_i = A \cap \{x: \dim(\text{image } Df(x)) \leq i\}.$$

Then the following results are true:

(i) The set Y_i is H^i-almost countably i-rectifiable.

(ii) For H^i-almost all $y \in Y_i$, $A \cap f^{-1}\{y\}$ is H^{m-i}-almost countably $(m-i)$-rectifiable.

(iii) $L^m(f^{-1}(Y_i) \sim X_i) = 0$.

Then the last result, relating Hausdorff and classical areas, becomes

$$\int_{Y_i} H^{m-i}(A \cap f^{-1}\{y\}) \, dH^i y = \int_{f^{-1}(Y^i)} J_i f \, dL^m(x).$$

Suppose now we take integers $r \in [1, \infty)$ and let $f: \mathbb{R}^m \to \mathbb{R}^n$ be of class r. Then X_i is the set of points where the differential has rank at most i.

What can be said of $f(X_i)$, precisely, for which s we have $H^s(f(X_i)) = 0$ or > 0? The answer is that $H^s(f(X_i)) = 0$ for $s = i + (m-i)/r$, and for $s < i + (m-i)/r$ we may have

$$H^s(f(X_i)) > 0.$$

For $i = m$ we may say this is classical. For $i = 0$ we have the results of A. P. Morse, and in the case $0 < i < m$ there are partial results of A. Sard.

A purely measure theoretic result on which existence theorems of the calculus of variations are based is the following:

Suppose we are given a Radon measure μ over $\mathbb{R}^n$ and we ask about conditions under which we can express this measure as an integral over a "nice" set, i.e., write

$$\mu(S) = \int_S \lambda \, dH^m x \quad \text{for all Borel } S$$

where λ is a nonnegative function (density) and $\{x: \lambda(x) > 0\}$ is countably m-rectifiable.

The answer is that this holds if and only if

$$\limsup_{r \to 0^+} \frac{\mu(B(x, r))}{\alpha(m) \, r^m} > 0 \qquad (1)$$

for μ-almost all $x \in \mathbb{R}^n$, where $\alpha(m)$ is the volume of m-dimensional ball; and

$$I^m(S) = 0 \implies \mu(S) = 0. \qquad (2)$$

2. CURRENTS

Now we turn to the generalized varieties, based on the above concepts, which prove to be useful in the calculus of variations. They were introduced in [2].

Let K be a compact set, i an integer, and φ a differential form. We put

$$\nu_K^i(\varphi) = \sup\{|D^j \varphi(x)| : j \leq i \text{ and } x \in K\}$$

and consider linear functionals T on the space of these forms for which for every K there exists an integer i and a constant $C < +\infty$ such that $T(\varphi) \leq C \nu_K^i(\varphi)$ whenever spt $\varphi \subset K$.

Call these functionals *currents*. Now define $\partial T: \varphi \to T(d\varphi)$ for all i − 1 forms φ as the boundary of the current T. The most important currents are those represented by integration.

CALCULUS OF VARIATIONS

To obtain those take $i = 0$ for all K; associated with T, there we define a function $\|T\|$ on f continuous and nonnegative, by the formula

$$\|T\|(f) = \sup\{T(\varphi) : \|\varphi\| \leq f,$$

for continuous and nonnegative $f\}$.

Then, $\|T\|(f) < +\infty$ for all f implies the existence of a Radon measure on $\mathbb{R}^n$, which we represent by $\|T\|$.

Now the general representation of T is

$$T(\varphi) = \int_{\mathbb{R}^n} <\vec{T}(x), \varphi(x)>d\|T\|(x)$$

for all admissible φ, where in the integrand there is an m vector $\vec{T}(x)$ geometrically associated with x in $\|T\|$-almost every $x \in \mathbb{R}^n$.

Let us examine the geometric role of the m vector $\vec{T}(x)$. In fact, currents are objects in the dual space of differential forms on $\mathbb{R}^n$. A very important class of currents are the *rectifiable* currents. These are currents for which conditions (1) and (2) of Sec. 1 hold, with $\mu = \|T\|$; λ is integer valued, and $\vec{T}(x)$ is an m vector which exists for μ-almost all $x \in \mathbb{R}^n$. In fact this allows us to associate with the object T in the dual space a geometric m vector in $\mathbb{R}^n$, which we call the tangent to T at x. This is, in a sense, the counterpart of Lebesgue's theorem on the existence, almost everywhere, of the derivative of a function of bounded variation.

It can be shown that rectifiable currents have approximation properties by more geometric objects. We use the term k-*polyhedral chain* to describe a current which is a finite linear consination of k-cells, i.e., currents of the type $\int_A \varphi$, where φ are k-forms on $\mathbb{R}^n$.

A Lipschitzian map $f: \mathbb{R}^m \to \mathbb{R}^n$ induces, in a classical way, a map $f_\#$ on the space of currents. We call Q an *integral Lipschitzian chain* in $\mathbb{R}^n$ if it is the image of polyhedral chain by a Lipschitzian function. Then rectifiable currents are $\|T\|$

approximates of a integral Lipschitzian chains. We refer to [3] for details.

3. VARIATIONAL PROBLEMS

Now if we represent the space of k-vectors of $\mathbb{R}^n$ by $A_k(\mathbb{R}^n)$, we may consider functions

$$F: \mathbb{R}^n \times \Lambda_k(\mathbb{R}^n) \to \mathbb{R}$$

which are continuous, nonnegative and positively homogeneous in the second argument, as classically assumed in the Calculus of Variations. Under quite general conditions of convexity on the integrand F, we may show that the functional

$$<T, F> = \int_{spt\ T} F(x, \vec{T}(x))\ dH^k(x)$$

is lower semicontinuous for rectifiable currents, with respect to convergence associated with Whitney's norm (see [4]):

$$W(T) = \inf\{M(U) + M(V): T = U + V,$$

$$U, V\ being\ k, (k+1)\ currents\}$$

where $M(U) = \sup\{U(\varphi): \varphi$ is a k-form in the unity ball$\}$.

For this semicontinuity theorem see, for example, [5].

Semicontinuity, associated with compactness properties, is enough to ensure the existence of a minimum for the functionals considered.

In fact, compactness is obtained in a fairly large class of currents.

Call T an *integral* current if both T and ∂T are rectifiable. Then we may introduce the norm of T as

CALCULUS OF VARIATIONS

$$N(T) = M(T) + M(\partial T)$$

and call T *normal* if $N(T) < +\infty$. It was shown in [2] that the class of integral currents with bounded norm is compact, which allow for existence theorems.

But while normal currents impose somewhat stringent conditions on the boundary of admissible currents, we may weaken these restrictions by considering the M closure of the set of normal currents. To do this, we need to assure rectifiability in this new class.

Let us characterize these currents in the following way. Given $\varepsilon \geq 0$ and a current T, consider the number

$$N^*(\varepsilon, T) = \inf\{M(\partial Q): M(T - Q) \leq \varepsilon\}.$$

$N^*(\varepsilon, T)$ is a function with values in $(0, +\infty)$, nonincreasing as a function of ε for any T. Following [6] we call *quasi-normal* those currents T for which $M(T) < +\infty$ and $N^*(\varepsilon, T) < +\infty$ for every $\varepsilon > 0$.

Let T be a k-dimensional quasi-normal current and consider the function $f: \mathbb{R}^n \to \mathbb{R}^k$. Then we say that T is f-*rectifiable* if, for every Borel set B, we have $f_\#(T \cap B)$ rectifiable. Then we have the following result:

THEOREM. Let T be a k-dimensional quasi-normal current in $\mathbb{R}^n$ and let Π_i, $i = 1, 2, \ldots, \binom{n}{k}$ be the projections of $\mathbb{R}^n$ on coordinate k space, and suppose T is Π_i-rectifiable. Then T is rectifiable.

For $k = n$, the proof is trivial. Then we assume the theorem true for $n - k = p$ and proceed by induction. To apply the induction process we need an additional hypothesis.

LEMMA. Let $\Pi: (x_1, x_2, \ldots, x_{n+1}) \in \mathbb{R}^{n+1} \to (x_1, x_2, \ldots, x_n) \in \mathbb{R}^n$, let the n-dimensional quasi-normal current T be Π-rectifiable, and assume there exists a number p such that

$$p(T \wedge dx_1 \wedge \cdots dx) \geq \|T\|.$$

Assume also that $\Pi_\# T$ is the elementary current associated by integration over the domain with the compact $K \subset \mathbb{R}^n$ with positive measure, and

supp $T = \{(y, f(y)): y \in K$, f continuous real valued in $K\}$.

Then there exists a compact $K' \subset K$, with positive measure, such that f is Lipschitzian in K'.

The proof of this is quite elaborate and depends essentially on a modified slicing process.

Then the main result of [5] can be carried to this larger class of currents, allowing for fairly general existence theorems. Details will appear elsewhere.

<u>Note added in proof</u>: Properties of quasi-normal currents are given in the dissertation "Correntes σ-normais," presented by João Rodrigues Filho at the Instituto Tecnológico de Aeronáutica, São José dos Campos, S. P., Brazil (October 1975).

REFERENCES

1. H. Federer, The (φ, k) rectifiable subsets of n-space, *Trans. Amer. Math. Soc.*, *62* (1947), 114-192.
2. H. Federer and W. M. Fleming, Normal and Integral Currents, *Ann. Math.*, *72* (1960), 458-520.
3. H. Federer, *Geometric Measure Theory*, Springer-Verlag, New York, 1969.
4. H. Whitney, *Geometric Integration Theory*, Princeton Univ. Press, Princeton, 1957.
5. U. D'Ambrosio, Semicontinuity of multiple integrals of the calculus of variations in parametric form, *Bull. Soc. Math. France*, *95* (1967), 357-384.
6. A. Chiffi, Correnti quasi-normali, *Ann. Scuola Nor. Sup. Pisa*, Ser. III, *20* (1966), 1-29.

ON THE LOCAL SOLVABILITY OF LINEAR PARTIAL DIFFERENTIAL OPERATORS[*]

Antonio Gilioli

Instituto de Matemática e Estatística
Universidade de São Paulo
São Paulo, Brazil

1. INTRODUCTION

As is customary, by a linear partial differential operator of degree $\leq m$ on an open set O of $\mathbb{R}^n$ we mean an operator of the form

$$P = \sum_{|\alpha| \leq m} a_\alpha(x) D^\alpha,$$

where

$$\alpha = (\alpha_1, \ldots, \alpha_n) \in \mathbb{N}^n, \quad |\alpha| = \alpha_1 + \cdots + \alpha_n,$$

$$D^\alpha = \frac{\partial^{|\alpha|}}{\partial x_1^{\alpha_1} \cdots \partial x_n^{\alpha_n}}, \quad a_\alpha \in C^\infty(O).$$

We recall the usual concept of local solvability:

---

[*] This work was partially supported by BNDE and FINEP.

DEFINITION 1. If $x_0 \in \mathcal{O}$, we say that P is locally solvable at x_0 iff there is an open set $\Omega \subset \mathcal{O}$, with $x_0 \in \Omega$, such that, for every $f \in C_c^\infty(\Omega)$, there is a distribution $u \in \mathcal{D}'(\Omega)$ such that $Pu = f$ on Ω.

We are mainly interested in two other (stronger) concepts of local solvability, as given by:

DEFINITION 2. If $x_0 \in \mathcal{O}$, we say that P is C^∞ locally solvable ($\mathcal{D}'$ locally solvable) at x_0 iff there is an open set $\Omega \subset \mathcal{O}$ with $x_0 \in \Omega$ such that, for every $f \in C_c^\infty(\Omega)$ $[f \in \mathcal{E}'(\Omega)]$, there is $u \in C^\infty(\Omega)$ $[u \in \mathcal{D}'(\Omega)]$ such that $Pu = f$ on Ω.

We shall give characterizations of the C^∞ and $\mathcal{D}'$ local solvability of a differential operator P. [As a matter of fact, we shall only use the fact that P is a linear continuous operator from $C_c^\infty(\Omega)$, $C^\infty(\Omega)$, $\mathcal{E}'(\Omega)$, and $\mathcal{D}'(\Omega)$, respectively, into themselves.]

Let us call $C^\infty(\overline{\Omega})$ the space of functions on Ω which can be extended to functions in $C^\infty(\mathbb{R}^n)$. Similarly, let $\mathcal{D}'(\overline{\Omega})$ be the space of distributions in Ω which can be extended to distributions in $\mathcal{D}'(\mathbb{R}^n)$.

It is easy to see (shrinking the Ω of Definition 2, if necessary) that Definition 2 is equivalent to the following:

DEFINITION 2'. If $x_0 \in \mathcal{O}$, we say that P is C^∞ ($\mathcal{D}'$) locally solvable at x_0 iff there is an open set $\Omega \subset \mathcal{O}$ with $x_0 \in \Omega$ such that, for every $f \in C^\infty(\overline{\Omega})$ $[f \in \mathcal{D}'(\overline{\Omega})]$, there is $u \in C^\infty(\Omega)$ $[u \in \mathcal{D}'(\overline{\Omega})]$ such that $Pu = f$ in Ω.

Definition 2' has the advantage that P is now defined and valued on the same space. Moreover, we may endow $C^\infty(\overline{\Omega})$ and

LOCAL SOLVABILITY

$\mathcal{D}'(\overline{\Omega})$ with topologies which are more well behaved than those of, for instance, $C_c^\infty(\Omega)$ and $\mathcal{D}'(\Omega)$.

In Sec. 2 we briefly review the definition of these and similar spaces, and their main properties. Almost all the proofs may be found in [1] (some in [2]). Where we could find no reference, we give the idea of the proof.

In Sec. 3 we use these spaces to characterize C^∞ and $\mathcal{D}'$ local solvability.

2. SOME SPACES OF FUNCTIONS AND DISTRIBUTIONS

In all this section, Ω will be a bounded nonempty open subset of $\mathbb{R}^n$ such that $\overline{\Omega}$ is an infinitely differentiable manifold with boundary $\partial\Omega$.

We define:

for $k \in \mathbb{N}$: $C^k(\overline{\Omega}) = \{f \in C^k(\Omega) \mid \forall \alpha \in \mathbb{N}^n$ with $|\alpha| \leq k$,

$\exists g_\alpha \in C(\overline{\Omega})$ such that $D^\alpha f = g_\alpha$ on $\Omega\}$,

$C^\infty(\overline{\Omega}) = \{f \in C^\infty(\Omega) \mid \forall \alpha \in \mathbb{N}^n, \exists g_\alpha \in C(\overline{\Omega})$

such that $D^\alpha f = g_\alpha$ on $\Omega\}$.

With the norm $\|f\|_k = \sup_{|\alpha| \leq k} \sup_{x \in \Omega} |D^\alpha f(x)|$, $C^k(\overline{\Omega})$ becomes a Banach space. The inclusions $i: C^{k+1}(\overline{\Omega}) \to C^k(\overline{\Omega})$ are compact operators and $C^\infty(\overline{\Omega})$ is a Fréchet-Schwartz (FS) space, the projective limit of the spaces $C^k(\overline{\Omega})$. We recall that FS spaces are Montel, hence reflexive.

Similarly, we consider the Sobolev spaces:

for $m \in \mathbb{N}$: $H^m(\Omega) = \{u \in \mathcal{D}'(\Omega) \mid D^\alpha u \in L^2(\Omega), \forall \alpha \in \mathbb{N}^n$

with $|\alpha| \leq m\}$,

$$H^\infty(\Omega) = \{u \in \mathcal{D}'(\Omega) \mid D^\alpha u \in L^2(\Omega), \forall \alpha \in \mathbb{N}^n\}.$$

With the inner product $(u, v) = \Sigma_{|\alpha| \le m} \int_\Omega D^\alpha u \, \overline{D^\alpha v}$, $H^m(\Omega)$ becomes a Hilbert space, whose norm will also be denoted by $\| \ \|_m$. The inclusions $i: H^{m+1}(\Omega) \to H^m(\Omega)$ are compact operators and $H^\infty(\Omega)$ is an FS space, the projective limit of $H^m(\Omega)$. Since the projective systems of $C^k(\overline{\Omega})$ ($k \in \mathbb{N}$) and $H^m(\Omega)$ ($m \in \mathbb{N}$) are equivalent (in the sense of [2]), it follows that $H^\infty(\Omega) = C^\infty(\overline{\Omega})$.

Moreover, we have (see [1] for the first equality):

$$(m \in \mathbb{N}) \quad H^m(\Omega) = \{f \in L^2(\Omega) \mid \exists \, g \in H^m(\mathbb{R}^n) \text{ such that } g = f \text{ on } \Omega\},$$

$$(k \in \mathbb{N}) \quad C^k(\overline{\Omega}) = \{f \in C^k(\Omega) \mid \exists \, g \in C^k(\mathbb{R}^n) \text{ such that } g = f \text{ on } \Omega\},$$

$$C^\infty(\overline{\Omega}) = \{f \in C^\infty(\Omega) \mid \exists \, g \in C^\infty(\mathbb{R}^n) \text{ such that } g = f \text{ on } \Omega\}.$$

The proof of the last equality relies on the following fact (which can be proved similarly to the proof of Borel's theorem, in [3]): The map $C^\infty(\mathbb{R}^n) \to C^\infty(\mathbb{R}^{n-1})^{\mathbb{N}}$, which sends each $\varphi \in C^\infty(\mathbb{R}^n)$ on $(\varphi_m)_{m \in \mathbb{N}}$, with $\varphi_m(x_1, \ldots, x_{n-1}) = (\partial^m \varphi / \partial x_n^m)(x_1, \ldots, x_{n-1}, 0)$, $\forall m \in \mathbb{N}$, is a surjective map.

The space $C^\infty(\overline{\Omega})$ is dense in all the spaces $C^k(\overline{\Omega})$ and $H^m(\Omega)$ (for $k \in \mathbb{N}$, $m \in \mathbb{N}$).

We now define

$$C_0^k(\overline{\Omega}) = \text{closure of } C_c^\infty(\Omega) \text{ in } C^k(\overline{\Omega}) \text{ (for } k \in \mathbb{N} \text{ or } k = \infty),$$

$$H_0^m(\Omega) = \text{closure of } C_c^\infty(\Omega) \text{ in } H^m(\Omega) \text{ (for } m \in \mathbb{N}).$$

Of course $C_0^\infty(\overline{\Omega})$ is an FS space, the projective limit of $C_0^k(\overline{\Omega})$, $k \in \mathbb{N}$, or of $H_0^m(\Omega)$, $m \in \mathbb{N}$. One can see that

LOCAL SOLVABILITY

$$C_0^k(\overline{\Omega}) = \{f \in C^k(\overline{\Omega}) \mid (D^\alpha f)(x) = 0 \;\; \forall \; x \in \partial\Omega, \; \forall \; \alpha \in \mathbb{N}^n$$

$$\text{with } |\alpha| \leq k\} \quad \text{for} \quad k \in \mathbb{N},$$

$$C_0^\infty(\overline{\Omega}) = \{f \in C^\infty(\overline{\Omega}) \mid (D^\alpha f)(x) = 0 \;\; \forall \; x \in \partial\Omega, \; \forall \; \alpha \in \mathbb{N}^n\}.$$

If we call $C_c^k(\overline{\Omega}) = \{f \in C^k(\mathbb{R}^n) \mid \operatorname{supp} u \subset \overline{\Omega}\}$ (for $k \in \mathbb{N}$ or $k = \infty$), then the restriction map $r: C_c^\infty(\overline{\Omega}) \to C_0^k(\overline{\Omega})$ which sends f in the restriction of f to Ω is an isometry onto $C_0^k(\overline{\Omega})$, for $k \in \mathbb{N}$, and a topological isomorphism for $k = \infty$. The same applies to $H_c^m(\overline{\Omega})$ (with the evident definition) and $H_0^m(\Omega)$, for $m \in \mathbb{N}$. Hence one usually identifies these spaces.

We shall denote by $H^{-m}(\Omega)$, $H^{-m}(\overline{\Omega})$, $\mathcal{D}'^k(\overline{\Omega})$, $E'^k(\overline{\Omega})$, $\mathcal{D}'(\overline{\Omega}) = \mathcal{D}'^\infty(\overline{\Omega})$, $E'(\overline{\Omega}) = E'^\infty(\overline{\Omega})$, respectively the strong duals of $H_0^m(\Omega)$, $H^m(\Omega)$, $C_0^k(\overline{\Omega})$, $C^k(\overline{\Omega})$, $C_0^\infty(\overline{\Omega})$, $C^\infty(\overline{\Omega})$ for $m \in \mathbb{N}$ and $k \in \mathbb{N}$. Of course the first two types of spaces are Hilbert spaces, the next two kinds are Banach spaces, and the last two are Silva spaces. $\mathcal{D}'(\overline{\Omega})$ is the inductive limit of the spaces $\mathcal{D}'^k(\overline{\Omega})$, $k \in \mathbb{N}$, or of $H^{-m}(\Omega)$, $m \in \mathbb{N}$; $E'(\overline{\Omega})$ is the inductive limit of the spaces $E'^k(\overline{\Omega})$, $k \in \mathbb{N}$, or of $H^{-m}(\overline{\Omega})$, $m \in \mathbb{N}$.

The transpose of the embeddings $i: C_0^k(\overline{\Omega}) \to C^k(\overline{\Omega})$ [$i: H_0^m(\Omega) \to H^m(\Omega)$] are surjective (by the Hahn-Banach theorem), hence also open maps $r: E'^k(\overline{\Omega}) \to \mathcal{D}'^k(\overline{\Omega})$ (for $k \in \mathbb{N}$ or $k = \infty$) [$r: H^{-m}(\overline{\Omega}) \to H^{-m}(\Omega)$].

Since $i: C_c^k(\Omega) \to C_0^k(\overline{\Omega})$ has dense image, it follows that its transpose is an injection $i: \mathcal{D}'^k(\overline{\Omega}) \to \mathcal{D}'^k(\Omega)$ for $k \in \mathbb{N}$ or $k = \infty$. Similar results are true for the spaces H^m.

Since $i: \mathcal{D}'^k(\overline{\Omega}) \to \mathcal{D}'^k(\Omega)$ is injective, we may identify (as sets) $\mathcal{D}'^k(\overline{\Omega})$ with a subspace of $\mathcal{D}'^k(\Omega)$, in the following way:

$$\mathcal{D}'^k(\overline{\Omega}) = \{u \in \mathcal{D}'^k(\Omega) \mid \exists \; v \in \mathcal{D}'^k(\mathbb{R}^n) \text{ such that } v = u \text{ on } \Omega\}$$

$$\text{for} \quad k \in \mathbb{N} \quad \text{or} \quad k = \infty.$$

Similarly one can make the following identification:

$$E'^k(\overline{\Omega}) = \{u \in \mathcal{D}'^k(\mathbb{R}^n) \mid \text{supp } u \subset \overline{\Omega}\} \quad \text{for} \quad k \in \mathbb{N} \quad \text{or} \quad k = \infty.$$

One also sees that

$$H^{-m}(\overline{\Omega}) = \{u \in H^{-m}(\mathbb{R}^n) \mid \text{supp } u \subset \overline{\Omega}\} \quad \text{for} \quad m \in \mathbb{N},$$

$C_c^\infty(\Omega)$ is dense in $\mathcal{D}'(\overline{\Omega})$ and $E'(\overline{\Omega})$.

Of course if P is a partial differential operator on $\mathcal{O}$, and $\overline{\Omega} \subset \mathcal{O}$, then $P: C^\infty(\overline{\Omega}) \to C^\infty(\overline{\Omega})$, $P: C_0^\infty(\overline{\Omega}) \to C_0^\infty(\overline{\Omega})$, $P: \mathcal{D}'(\overline{\Omega}) \to \mathcal{D}'(\overline{\Omega})$ and $P: E'(\overline{\Omega}) \to E'(\overline{\Omega})$ are all continuous maps. The same happens if the coefficients of P belong to $C^\infty(\overline{\Omega})$.

3. CHARACTERIZATION OF C^∞ AND $\mathcal{D}'$ LOCAL SOLVABILITY

We still assume throughout this section that Ω is a bounded nonempty open subset of $\mathbb{R}^n$, with smooth boundary.

We shall use the following theorem on Functional Analysis:

Let E and F be Fréchet spaces, $T: E \to F$ a linear continuous map. Then the following conditions are equivalent:

(a) T is surjective.

(a') T is a surjective homomorphism.

(b) $^tT: F' \to E'$ is injective and its image is weakly closed. (If E is also reflexive, then weakly closed is equivalent to strongly closed.)

(c) (If E is, for instance, FS) $^tT: F' \to E'$ is an injective homomorphism.

(d) If $\mathcal{B}$ and $\mathcal{B}'$ are bases of continuous seminorms on E and F, respectively, then, $\forall\, p \in \mathcal{B}$, $\exists\, q \in \mathcal{B}'$ and $\exists\, c > 0$

such that, $\forall\, y' \in F'$,

$${}^t T(y') \in (\hat{E}_p)' \implies y' \in (\hat{F}_q)' \text{ and } \|y'\|_q \leq c \|{}^t T(y')\|_p.$$

By ${}^t T$ we denote the transpose of T, and by E' the strong dual of E.

We say that T: E → F is a homomorphism if T is linear continuous and T: E → T(E) is an open map.

If E is a locally convex vector space and p a continuous seminorm on E, we call (E, p) the space E, endowed with the single seminorm p, which is a seminormed space. Then the quotient of (E, p) by the kernel of p is a normed space, which we denote E_p, and by $\hat{E}_p$ we denote the completion of E_p. Of course $\hat{E}_p$ is a Banach space, and we represent by $\|\ \|_p$ the dual norm on $(\hat{E}_p)'$.

The proof that (a) and (b) are equivalent may be found in [4], and the equivalence of (a) and (d) in [3].

Applying this theorem to the operator P: $C^\infty(\overline{\Omega}) \to C^\infty(\overline{\Omega})$, we get at once:

THEOREM 1. The following conditions are equivalent (Ω bounded, $\partial\Omega$ smooth):

(1) P: $C^\infty(\overline{\Omega}) \to C^\infty(\overline{\Omega})$ is surjective.

(2) ${}^t P: E'(\overline{\Omega}) \to E'(\overline{\Omega})$ is injective, with closed image.

(3) ${}^t P: E'(\overline{\Omega}) \to E'(\overline{\Omega})$ is an injective homomorphism.

(4) $\forall\, m_1 \in \mathbb{N}$, $\exists\, m_2 \in \mathbb{N}$ and $c > 0$ such that, $\forall\, u \in E'(\overline{\Omega})$,

$${}^t Pu \in H^{-m_1}(\overline{\Omega}) \implies u \in H^{-m_2}(\overline{\Omega}) \text{ and } \|u\|_{-m_2} \leq c \|{}^t Pu\|_{-m_1}.$$

One may give another equivalent statement, analogous to (4), using the norms of $E'^k(\overline{\Omega})$, instead of Sobolev's norms.

We shall also use the following fact, which is not difficult to verify:

Let E and F be the strong duals of Fréchet reflexive spaces, $T: E \to F$ a linear continuous map. Then the following are equivalent:

(a) $T: E \to F$ is surjective.
(a') $T: E \to F$ is a surjective homomorphism.
(b) ${}^tT: F' \to E'$ is injective, with closed image.
(c) ${}^tT: F' \to E'$ is an injective homomorphism.
(d) If B and B' are bases of continuous seminorms over F' and E', respectively, then, $\forall\, p \in B$, $\exists\, q \in B'$, and $c > 0$ such that, $\forall\, x' \in F'$, $p(x') \leq c\, q({}^tT(x'))$.

Applying this theorem to the operator $P: \mathcal{D}'(\overline{\Omega}) \to \mathcal{D}'(\overline{\Omega})$, we get at once:

THEOREM 2. The following conditions are equivalent:

(1) $P: \mathcal{D}'(\overline{\Omega}) \to \mathcal{D}'(\overline{\Omega})$ is surjective (or a surjective homomorphism).
(2) ${}^tP: C_0^\infty(\overline{\Omega}) \to C_0^\infty(\overline{\Omega})$ is injective, with closed image.
(3) ${}^tP: C_0^\infty(\overline{\Omega}) \to C_0^\infty(\overline{\Omega})$ is an injective homomorphism.
(4) $\forall\, m_1 \in \mathbb{N}$, $\exists\, m_2 \in \mathbb{N}$ and $c > 0$ such that, $\forall\, \varphi \in C_0^\infty(\overline{\Omega})$,

$$\|\varphi\|_{m_1} \leq c \|{}^tP\varphi\|_{m_2}.$$

We end with some applications of these theorems.

PROPOSITION 1. If $P: C^\infty(\overline{\Omega}) \to C^\infty(\overline{\Omega})$ is surjective, then ${}^tP: \mathcal{D}'(\overline{\Omega}) \to \mathcal{D}'(\overline{\Omega})$ is a homomorphism with closed image.

LOCAL SOLVABILITY

Proof. By Theorem 1, we know that ${}^tP: E'(\overline{\Omega}) \to E'(\overline{\Omega})$ is an injective homomorphism with closed image. Since closed subspaces of Silva spaces are also Silva, it follows that ${}^tP(E'(\overline{\Omega}))$ is a Silva space. Let us call by A the operator ${}^tP: \mathcal{D}'(\overline{\Omega}) \to \mathcal{D}'(\overline{\Omega})$ and $r: E'(\overline{\Omega}) \to \mathcal{D}'(\overline{\Omega})$ the canonical surjective homomorphism. Since r is surjective, we have $A(\mathcal{D}'(\overline{\Omega})) = r({}^tP(E'(\overline{\Omega})))$. Since $r: E'(\overline{\Omega}) \to \mathcal{D}'(\overline{\Omega})$ is a surjective homomorphism, it follows that $r: {}^tP(E'(\overline{\Omega})) \to A(\mathcal{D}'(\overline{\Omega}))$ is also a homomorphism, which is surjective. Since $A(\mathcal{D}'(\overline{\Omega}))$ is the homomorphic image of the Silva space ${}^tP(E'(\overline{\Omega}))$, we conclude that $A(\mathcal{D}'(\overline{\Omega}))$ is also Silva, hence complete and therefore closed in $\mathcal{D}'(\overline{\Omega})$. Finally, since $A: \mathcal{D}'(\overline{\Omega}) \to A(\mathcal{D}'(\overline{\Omega}))$ is surjective continuous between Silva spaces, it follows that it is open.

COROLLARY. If $P: C^\infty(\overline{\Omega}) \to C^\infty(\overline{\Omega})$ is surjective, then the following conditions are equivalent:

(a) ${}^tP: \mathcal{D}'(\overline{\Omega}) \to \mathcal{D}'(\overline{\Omega})$ is surjective.

(b) ${}^tP: \mathcal{D}'(\overline{\Omega}) \to \mathcal{D}'(\overline{\Omega})$ has a dense image.

(c) $P: C_0^\infty(\overline{\Omega}) \to C_0^\infty(\overline{\Omega})$ is injective.

PROPOSITION 2. If $P: C^\infty(\overline{\Omega}) \to C^\infty(\overline{\Omega})$ is surjective and satisfies the condition

(H) $\quad\quad \forall\, u \in E'(\overline{\Omega}),\ {}^tPu \in C_0^\infty(\overline{\Omega}) \implies u \in C_0^\infty(\overline{\Omega}),$

then $P: \mathcal{D}'(\overline{\Omega}) \to \mathcal{D}'(\overline{\Omega})$ is surjective.

Proof. From Theorem 1, ${}^tP: E'(\overline{\Omega}) \to E'(\overline{\Omega})$ is injective with closed image, hence ${}^tP: C_0^\infty(\overline{\Omega}) \to C_0^\infty(\overline{\Omega})$ is injective. We have to prove that ${}^tP(C_0^\infty(\overline{\Omega}))$ is closed in $C_0^\infty(\overline{\Omega})$. Let f_n ($n \in \mathbb{N}$) and g belong to $C_0^\infty(\overline{\Omega})$, such that ${}^tPf_n$ converges to g in $C_0^\infty(\overline{\Omega})$.

Then, the convergence also holds in $E'(\overline{\Omega})$, and since ${}^tP(E'(\overline{\Omega}))$ is closed in $E'(\overline{\Omega})$, there is $u \in E'(\overline{\Omega})$ such that ${}^tPu = g$. By (H), we get $u \in C_0^\infty(\overline{\Omega})$.

Remark 1. One should not confuse condition (H) with hypoellipticity. Condition (H) is a very weak one (for sufficiently small open sets). For instance, it holds for all differential operators P with constant coefficients, or more generally, of constant strength (see [5]). The essential point is that in (H) we assume that u has compact support, while there is no such restriction in the definition of hypoellipticity.

Remark 2. It is possible to verify that the condition

$$\forall\, u \in E'(\overline{\Omega}),\ {}^tPu \in H^{-m_1}(\overline{\Omega}) \implies u \in H^{-m_2}(\overline{\Omega})$$

is equivalent to the following:

$$\forall\, m \in \mathbb{Z}\ (\text{near } -\infty),\ \exists\, C_m \text{ such that } \|u\|_{-m_2} \leq C_m(\|u\|_m + \|{}^tPu\|_{-m_1}),\ \text{for all } u \in H^m(\overline{\Omega}) \text{ such that } {}^tPu \in H^{-m_1}(\overline{\Omega})$$

Hence, apparently it is essential to keep the inequality of condition (4) in Theorem 1.

Remark 3. If P is hypoelliptic on Ω, and $x_0 \in \Omega$, it is possible to verify that, for each $m_1 \in \mathbb{N}$ given, there is a sufficiently small open neighborhood Ω_{m_1} of x_0 ($\Omega_{m_1} \subset \Omega$), and $m_2 \in \mathbb{N}$ such that inequality (4) of Theorem 2 holds for all $\varphi \in C_0^\infty(\overline{\Omega_{m_1}})$. Nevertheless, it is not clear whether one can choose a *fixed* open neighborhood Ω' of x_0 such that all the inequalities hold in Ω'. If this were the case, then ${}^tP$ would be D' locally solvable at each point of Ω'.

We believe that, at least when $P(C^\infty(\overline{\Omega})) = C^\infty(\overline{\Omega})$ and $P(\mathcal{D}'(\overline{\Omega})) = \mathcal{D}'(\overline{\Omega})$, there should be some constant c (probably ≥ 0) such that, $\forall\, m_1 \in \mathbb{N}$, $\exists\, C > 0$ such that $\|u\|_{m_1+c} \leq C\|{}^tPu\|_{m_1}$, $\forall\, u \in E'(\overline{\Omega})$ with ${}^tPu \in H_0^{m_1}(\Omega)$ if $m_1 \geq 0$, or with $u \in H^{m_1}(\overline{\Omega})$ if $m_1 < 0$. In fact, observe that we did not use the fact that, if P is of degree $\leq m$, then $P: H^{m_1}(\Omega) \to H^{m_1-m}(\Omega)$ is continuous. The assumptions on the operator T in these facts about Functional Analysis were very much less than what we know about differential operators.

REFERENCES

1. J. L. Lions and E. Magenes, *Problèmes aux Limites non homogènes et Applications*, vol. 1, Dunod, Paris, 1968.

2. H. Komatsu, Projective and injective limits of weakly compact sequences of locally convex spaces, *J. Math. Soc. Japan*, 19, N° 3 (1967), 366-383.

3. F. Trèves, *Locally Convex Spaces and Linear Partial Differential Operators*, Springer-Verlag, New York, 1967.

4. F. Trèves, *Topological Vector Spaces, Distributions and Kernels*, Academic Press, New York, 1967.

5. L. Hörmander, *Linear Partial Differential Operators*, 3rd Edition, Springer-Verlag, New York, 1969.

Note added in proof: Although facts (1), (2), and (3) below are implicit in the propositions, we state them explicitly, for the sake of clarity. They give three different sufficient conditions for the operator ^tP to be $\mathcal{D}'$ locally solvable. Facts (4), (5), and (6) are only references for the proof of the first three.

(1) If P is C^∞ locally solvable and ^tP is locally solvable, then ^tP is $\mathcal{D}'$ locally solvable.

(2) If P is hypoelliptic and locally solvable, then ^tP is $\mathcal{D}'$ locally solvable (cf. Remark 3).

(3) If P is hypoelliptic and ^tP is C^∞ locally solvable, then ^tP is $\mathcal{D}'$ locally solvable.

(4) If P is hypoelliptic and locally solvable, then P is C^∞ locally solvable.

(5) If P is hypoelliptic, then ^tP is locally solvable.

(6) If P is locally solvable at x_0, then $^tP: C_0^\infty(\overline{\Omega}) \to C_0^\infty(\overline{\Omega})$ is injective for a sufficiently small open set Ω containing x_0.

Proof. (4) is trivial; (5) is well known; (6) is a consequence of Hörmander's inequality (see Lemma 6.1.2 of [5]); (1) is a consequence of (6) and of the Corollary of Proposition 1; (2) is a particular case of (1), if we remember (4) and (5); (3) is a particular case of Proposition 2, if we exchange P for ^tP and ^tP for P.

ON STRONGLY NONLINEAR ELLIPTIC PROBLEMS

Peter Hess

Mathematisches Institut
Universität Zürich
Zürich, Switzerland

1. INTRODUCTION

The intensive development of the theory of monotone operators in reflexive Banach spaces started about a decade ago when it was realized that these mappings form a very powerful tool in discussing the solvability of boundary value problems for quasi-linear elliptic equations in divergence form

$$Au \equiv \sum_{|\alpha| \leq m} (-1)^{|\alpha|} D^{\alpha} A_{\alpha}(x, u, \ldots, D^m u) = f,$$

considered in a bounded domain $\Omega \subset \mathbb{R}^N$, provided the functions A_{α} satisfy some polynomial growth condition in u and its partial derivatives of order $\leq$ m. In connection with the decomposition of A into its top-order part ($|\alpha| = m$) and its lower-order terms, the concept of monotonicity was subsequently weakened, and various classes of nonlinear operators which we summarize as "mappings of monotone type" were introduced. For a survey see Browder [1].

If the growth of the A_α's is more than polynomial, it is customary to work with operators of monotone type in Orlicz-Sobolev spaces which may not be reflexive. For the study of questions arising in that context we refer to Gossez [2].

From point of view of applications, the question of solvability of boundary value problems for equations of the form

$$Au + g(u) = f \qquad (1.1)$$

is of particular importance. Here A denotes a linear or quasi-linear elliptic differential operator (with coefficient functions A_α subject to some polynomial growth restriction), while no growth assumption is imposed on the function g. Operators of the form $A + g$ are called strongly nonlinear. The pecularity of these problems is that it is still possible to work within the framework of reflexive Sobolev spaces. However, the abstract mappings induced are now in general neither everywhere defined nor bounded. After some pioneering work by Browder [3] and Ton [4], the author established in [5] a proper abstract setting for those problems by introducing the "mappings of monotone type with respect to two Banach spaces."

It is the purpose of this chapter to give a detailed treatment of the Neumann boundary value problem for an equation of the form of (1.1), provided A is a differential operator of second order. We note that this restriction of order is inessential if A is linear. We first prove an abstract existence theorem under a weak coerciveness assumption. This result provides us with a surprisingly simple method of approach to the strongly nonlinear "semicoercive" Neumann boundary value problem. We partly follow the author's papers [5, 6], sharpening the results obtained therein in various respects.

Variational inequalities for strongly nonlinear elliptic operators are considered in [7-9]. Recently strongly nonlinear elliptic boundary value problems and associated variational inequalities have also been discussed in *unbounded* domains by Edmunds et al. [10] as well as Webb [11].

2. AN ABSTRACT EXISTENCE THEOREM

In the following let W, V denote two real, reflexive, and separable Banach spaces with norms $\|\cdot\|_W$ and $\|\cdot\|_V$, respectively, and assume $W \subset V$, with a continuous injection mapping of W into V. Let W^*, V^* be the spaces conjugate to W and V. Then $V^* \subset W^*$ in the sense that if $f|_W$ is the restriction of the functional $f \in V^*$ to W, then $f|_W \in W^*$. By (g, u) we denote the duality pairing, either between $g \in W^*$, $u \in W$, or between $g \in V^*$, $u \in V$. The symbols $\to$ and $\rightharpoonup$ describe strong and weak convergence, respectively; it is, however, always specified in which space convergence is understood.

DEFINITION 2.1. Let T be a mapping with domain $\mathcal{D}(T)$: $W \subset \mathcal{D}(T) \subset V$, and range contained in W^*. Then T is said to be *of type (M) with respect to* W, V, provided

(i) T is continuous from finite-dimensional subspaces of W to the weak topology on W^*, and

(ii) if $\{v_n\}$ is any sequence in W and $u \in V$, $g \in V^*$ elements such that $v_n \rightharpoonup u$ in V, $Tv_n \rightharpoonup g|_W$ in W^*, and $\limsup (Tv_n, v_n) \leq (g, u)$, then $u \in \mathcal{D}(T)$ and $Tu = g|_W$.

quasi-bounded if for any sequence $\{v_n\}$ in W which is bounded in the V norm, and for which $(Tv_n, v_n) \leq \text{const} \|v_n\|_V$ $\forall$ n, the boundedness in W^* of the sequence $\{Tv_n\}$ follows.

These notations are justified by the fact that if $W = V$, they coincide with the concepts usually employed.

Our basic existence result is

THEOREM 2.2. Let the mapping T be quasi-bounded and of type (M) with respect to W, V, and suppose for some $R > 0$,

$$(Tw, w) > 0 \quad \forall \, w \in W \text{ with } \|w\|_V = R.$$

Then there exists $u \in \mathcal{D}(T)$, $\|u\|_V \leq R$, such that

$$Tu = 0.$$

We note that if f is any given element in V^*, then together with T the mapping $T_f: \mathcal{D}(T) \to W^*$ defined by

$$T_f v = Tv - f|_W \quad [v \in \mathcal{D}(T)]$$

is quasi-bounded and of type (M) with respect to W, V. Thus, as a consequence of Theorem 2.2, the equation

$$Tu = f|_W$$

is solvable for *all* $f \in V^*$ provided T is *coercive*:

$$\|w\|_V^{-1} (Tw, w) \to +\infty \text{ as } \|w\|_V \to \infty, \quad w \in W.$$

<u>Proof of Theorem 2.2.</u> Let $\{W_n\}$ be an increasing sequence of finite-dimensional subspaces of W, whose union is dense in W. For each n let j_n denote the injection mapping of W_n into W and $j_n^*: W^* \to W_n^*$ the dual projection. The mapping $T_n = j_n^* T j_n : W_n \to W_n^*$ is then continuous, and $(T_n w, w) = (Tw, w) > 0 \; \forall \, w \in W_n$ with $\|w\|_V = R$. Since the W norm and the V norm are equivalent on the finite-dimensional space W_n, we infer by a standard argument (e.g., [1]) the existence of $v_n \in W_n$ with $\|v_n\|_V < R$ such that $T_n v_n = 0$.

By reflexivity of V we may assume (passing to a suitable subsequence) that $v_n \rightharpoonup u$ in V. Since

$$(Tv_n, v_n) = (T_n v_n, v_n) = 0 = (0, u),$$

the quasi-boundedness of T implies the boundedness in W^* of the

NONLINEAR ELLIPTIC PROBLEMS

sequence $\{Tv_n\}$. Let $w \in \bigcup_n W_n$. Then $w \in W_n$ for n large enough, and hence

$$(Tv_n, w) = (T_n v_n, w) = 0 = (0, w)$$

for those n. Since $\bigcup_n W_n$ is dense in W and $\|Tv_n\|_{W^*}$ is bounded, we conclude that $Tv_n \rightharpoonup 0$ in W^*. T being of type (M) with respect to W, V, it follows that $u \in \mathcal{D}(T)$ and $Tu = 0$.

3. A STRONGLY NONLINEAR ELLIPTIC BOUNDARY VALUE PROBLEM

Let $\Omega \subset \mathbb{R}^N$ ($N \geq 1$) be a bounded domain with smooth boundary Γ. We shall apply Theorem 2.2 in the discussion of solvability of the strongly nonlinear elliptic boundary value problem of Neumann type

(N)
$$Au + g(\cdot, u) = f \quad \text{in } \Omega$$
$$\frac{\partial u}{\partial n_a} = 0 \quad \text{on } \Gamma.$$

Here A:

$$Au = -\sum_{i=1}^{N} \frac{\partial}{\partial x_i} A_i(x, u, \text{grad } u)$$

denotes a (possibly nonlinear) differential operator giving rise to a "regular" boundary value problem (in particular, the A_i satisfy some polynomial growth condition in u and grad u), while $g: \Omega \times \mathbb{R} \to \mathbb{R}$ is a function which is only subject to some sign condition. $\partial u / \partial n_a$ is the outward conormal derivative associated to A by Green's formula

$$\int_\Omega Au \cdot v \, dx = a(u, v) - \int_\Gamma \frac{\partial u}{\partial n_a} v \, d\Gamma$$

[u, v in $C^1(\overline{\Omega})$], with the form a as defined below.

More precisely we assume

(AI) Each A_i ($i = 1, \ldots, N$) is a function from $\Omega \times \mathbb{R} \times \mathbb{R}^N$ to $\mathbb{R}$ satisfying the Caratheodory condition: $A_i(x, t, \xi)$ is measurable in $x \in \Omega$ for fixed $(t, \xi) \in \mathbb{R} \times \mathbb{R}^N$, and continuous in (t, ξ) for a.a. fixed x. Moreover there exist constants c and p: $1 < p < \infty$ such that

$$|A_i(x, t, \xi)| \leq c(1 + |t|^{p-1} + |\xi|^{p-1})$$

for $i = 1, \ldots, N$, a.a. $x \in \Omega$, and $\forall (t, \xi) \in \mathbb{R} \times \mathbb{R}^N$.

Condition (AI) implies that the (semilinear) form

$$a(u, v) = \sum_{i=1}^{N} \int_\Omega A_i(\cdot, u, \text{grad } u) \frac{\partial v}{\partial x_i} dx$$

is defined for all u, v in the space $V = W^{1,p}(\Omega)$. By standard results about Nemytskii operators (e.g., [12]), the mapping $A: V \to V^*$ given by

$$(Au, v) = a(u, v) \quad (u, v \text{ in } V) \tag{3.1}$$

is bounded and continuous. We further suppose

(AII) $\sum_{i=1}^{N} [A_i(x, t, \xi) - A_i(x, t, \xi')](\xi_i - \xi_i') \geq 0$ for a.a. $x \in \Omega$, $\forall t \in \mathbb{R}$, and $\forall \xi, \xi' \in \mathbb{R}^N$.

It is well known that then the mapping A defined by (3.1) is of *type* (M): for any sequence $\{u_n\}$ in V with $u_n \rightharpoonup u$ in V and $Au_n \rightharpoonup \chi$ in V^*, for which $\limsup(Au_n, u_n) \leq (\chi, u)$, it follows that $\chi = Au$ (e.g., Lions [13, Chapter 2.2]).

The conditions imposed on the function g are as follows:

(BI) $g: \Omega \times \mathbb{R} \to \mathbb{R}$ satisfies the Caratheodory condition, and there exists $T \geq 0$ such that

$$g(x, t) t \geq 0 \quad \forall \; |t| \geq T, \quad \text{a.a.} \quad x \in \Omega.$$

(BII) For any $\tau > 0$,

$$\int_\Omega \sup_{|t| \leq \tau} |g(x, t)| \, dx < \infty.$$

Note that no assumption is made restricting the growth of g as $|t| \to \infty$.

Let $W = W^{m,p}(\Omega)$, where $m > 1 + [N/p]$, so that $W \subset C(\overline{\Omega})$ by Sobolev's imbedding theorem. By Green's formula, if u is a solution of problem (N) and $w \in W$, then formally

$$a(u, w) + \int_\Omega g(\cdot, u) w \, dx = \int_\Omega f w \, dx.$$

The integral

$$b(u, w) = \int_\Omega g(\cdot, u) w \, dx,$$

however, makes sense for arbitrary $w \in W$ only if $g(\cdot, u)$ is integrable on Ω. We therefore introduce the set

$$V_1 = \{u \in V: g(\cdot, u) \in L^1(\Omega), \; g(\cdot, u) u \in L^1(\Omega)\}.$$

Thus $b(u, w)$ is well defined for $u \in V_1$, $w \in W$. Evidently $W \subset V_1 \subset V$.

We are led to the following:

DEFINITION 3.1. Given $f \in W^*$, a function u is said to be weak solution of problem (N) if

(i) $u \in V_1$,
(ii) $a(u, w) + b(u, w) = (f, w) \quad \forall \; w \in W$.

To fixed $u \in V_1$ there exists $u^* \in W^*$ such that

$$(u^*, w) = b(u, w) \quad \forall\, w \in W$$

(i.e., the linear functional $w \mapsto b(u, w)$ is bounded on W). The correspondence $u \mapsto u^*$ defines a mapping $B: V_1 \to W^*$ by $Bu = u^*$. Let the mapping $\tilde{A}: V \to W^*$ be given by $\tilde{A}u = Au|_W$ ($u \in V$), and set $T = \tilde{A} + B: \mathcal{D}(T) = V_1 \to W^*$.

PROPOSITION 3.2. Suppose conditions (AI, II) and (BI, II) hold. Then the mapping T defined above is quasi-bounded and of type (M) with respect to W, V.

Obviously a function u is weak solution of problem (N) iff $Tu = f$. A criterion for solvability of problem (N) is thus given by

THEOREM 3.3. Let conditions (AI, II) and (BI, II) be satisfied. Then problem (N) admits a weak solution for given $f \in V^*$ provided there exists $R > 0$ such that

$$a(w, w) + b(w, w) - (f, w) > 0 \quad \forall\, w \in W \text{ with } \|w\|_V = R.$$

The proof of Proposition 3.2 will be accomplished in Sec. 4. We now state, as a consequence of Theorem 3.3, a condition for solvability which is immediately applicable. Let us assume that A satisfies, in addition to (AI, II), condition

(AIII) There exists $\alpha > 0$ such that

$$\sum_{i=1}^{N} A_i(x, t, \xi)\, \xi_i \geq \alpha \sum_{i=1}^{N} |\xi_i|^p$$

for a.a. $x \in \Omega$, $\forall\, (t, \xi) \in \mathbb{R} \times \mathbb{R}^N$.

NONLINEAR ELLIPTIC PROBLEMS

For a.a. $x \in \Omega$ let further

$$g_+(x) = \liminf_{t \to +\infty} g(x, t),$$

$$g_-(x) = \limsup_{t \to -\infty} g(x, t). \tag{3.2}$$

COROLLARY 3.4. Suppose conditions (AI-III) and (BI, II) hold. Sufficient for the weak solvability of problem (N), for given $f \in L^{p'}(\Omega)$ $[p' = p/(p - 1)]$, is the validity of the inequalities

$$\int_\Omega g_- \, dx < \int_\Omega f \, dx < \int_\Omega g_+ \, dx. \tag{3.3}$$

(Since $g_\pm$ may attain the values $\pm\infty$ on sets with positive measure, the integrals on the left and right in (3.3) may be $-\infty$ and $+\infty$, respectively.)

If, in particular,

$$g_-(x) < g(x, t) < g_+(x) \quad \text{for a.a. } x \in \Omega, \ \forall \, t \in \mathbb{R},$$

condition (3.3) is easily seen to be also *necessary*. Corollary 3.4 is a result on a "semicoercive" problem, as they have been considered, by different methods, in the author's paper [14] and in [15].

Proof of Corollary 3.4. We show the existence of a constant $R > 0$ such that

$$a(w, w) + b(w, w) - (f, w) > 0 \quad \forall \, w \in W \text{ with } \|w\|_V = R.$$

The assertion then follows from Theorem 3.3.

Suppose, to the contrary, that there exists a sequence $\{w_n\}$ in W with $\|w_n\|_V \to \infty$ and

$$a(w_n, w_n) + b(w_n, w_n) - (f, w_n) \leq 0 \quad \forall\, n. \tag{3.4}$$

We decompose g into a sum $g = g_0 + h$ of functions both satisfying the Caratheodory condition, such that

$$g_0(x, t)\, t \geq 0 \quad \text{for a.a. } x \in \Omega,\ \forall\, t \in \mathbb{R},$$

$$h(x, t) = 0 \quad \forall\, |t| \geq T,\ \text{a.a. } x \in \Omega,\ \text{and} \tag{3.5}$$

$$\int_\Omega \sup_t |h(x, t)|\, dx < \infty.$$

Let $z_n = \|w_n\|_V^{-1} w_n$. Passing to subsequences we may assume that

$$z_n \rightharpoonup z \quad \text{in } V,$$

$$z_n \to z \quad \text{in } L^p(\Omega) \text{ as well as a.e. in } \Omega.$$

From (3.4) we infer that

$$\alpha \int_\Omega \sum \left|\frac{\partial z_n}{\partial x_i}\right|^p dx + \frac{1}{\|w_n\|_V^{p-1}} \\ \times \left[\int_\Omega g(\cdot, w_n)\, z_n\, dx - (f, z_n)\right] \leq 0 \quad \forall\, n. \tag{3.6}$$

Since

$$\int_\Omega g(\cdot, w_n)\, z_n\, dx \geq \int_\Omega h(\cdot, w_n)\, z_n\, dx \geq -\text{const.},$$

it follows that

$$\int_\Omega \sum \left|\frac{\partial z_n}{\partial x_i}\right|^p dx \to 0 \quad (n \to \infty)$$

and thus

$$z_n \to z = \text{const} \ (\neq 0) \text{ in } V.$$

By (3.6),

$$\int_\Omega g(\cdot, w_n) z_n \, dx - (f, z_n) \leq 0 \quad \forall \, n,$$

whence

$$\limsup \int_\Omega g(\cdot, w_n) z_n \, dx \leq (f, z) = \int_\Omega fz \, dx.$$

Suppose first that $z > 0$. Then $w_n = \|w_n\|_V \, z_n \to +\infty$ a.e. in Ω. By the Fatou lemma,

$$\int_\Omega g_+ \, z \, dx \leq \liminf \int_\Omega g_0(\cdot, w_n) \, z_n \, dx.$$

The Lebesgue theorem on dominated convergence further guarantees that

$$\int_\Omega h(\cdot, w_n) \, z_n \, dx \to 0 \quad \text{as } n \to \infty.$$

Together we obtain

$$\int_\Omega g_+ z \, dx \leq \int_\Omega fz \, dx$$

and thus

$$\int_\Omega g_+ \, dx \leq \int_\Omega f \, dx.$$

If $z < 0$, we similarly derive

$$\int_\Omega g_- \, dx \geq \int f \, dx.$$

In both cases we arrive at a contradiction to (3.3). Corollary 3.4 is thus proved.

COROLLARY 3.5. Suppose conditions (AI-III) hold, and let the function $g: \Omega \times \mathbb{R} \to \mathbb{R}$ satisfy the Caratheodory condition. Further, let $f \in L^{p'}(\Omega)$ be given. Suppose there exist constants $R_1 \leq 0 \leq R_2$ such that

$$\int_\Omega \sup_{R_1 \le t \le R_2} |g(x, t)| \, dx < \infty$$

and

$$g(x, R_1) \le f(x), \quad g(x, R_2) \ge f(x) \quad \text{for a.a. } x \in \Omega. \quad (3.7)$$

Then problem (N) admits a weak solution u with $R_1 \le u(x) \le R_2$ a.e. in Ω.

No assumption at all on the asymptotic behavior of $g(x, t)$ as $|t| \to \infty$ is made here.

Proof. We introduce the modified function

$$\begin{aligned} g_1(x, t) &= g(x, R_1) & t \le R_1 \\ &= g(x, t) & R_1 \le t \le R_2 \\ &= g(x, R_2) & t \ge R_2 \end{aligned}$$

$(x \in \Omega, t \in \mathbb{R})$. Further let $\beta: \mathbb{R} \to \mathbb{R}$ be the function given by

$$\begin{aligned} \beta(t) &= -(R_1 - t)^{p-1}, & t \le R_1; \\ &= 0, & R_1 \le t \le R_2; \\ &= (t - R_2)^{p-1}, & t \ge R_2. \end{aligned}$$

It is then readily seen that the function g_2:

$$g_2(x, t) = g_1(x, t) - f(x) + \beta(t)$$

$(x \in \Omega, t \in \mathbb{R})$ satisfies (BI, II). Since $g_2(x, t) \to \pm\infty$ as $t \to \pm\infty$, for a.a. $x \in \Omega$, Corollary 3.4 guarantees the existence of a function $u \in V$ such that

$$g_2(\cdot, u) \in L^1(\Omega), \quad g_2(\cdot, u) u \in L^1(\Omega), \tag{3.8}$$

and

$$a(u, w) + \int_\Omega g_2(\cdot, u) w \, dx = 0 \quad \forall w \in W. \tag{3.9}$$

We now show that u is actually an asserted solution of problem (N).

In order to accomplish this proof let $\gamma: \mathbb{R} \to \mathbb{R}$ denote a uniformly bounded, Lipschitz continuous and monotone increasing function such that

$$\gamma(t) = 0 \quad \text{iff} \quad R_1 \le t \le R_2.$$

Then $\gamma(u) \in V \cap L^\infty(\Omega)$. Thus there exists a suitable constant $M > 0$ and a sequence $\{h_k\}$ in $C^\infty(\overline{\Omega})$ with $|h_k(x)| \le M$ ($\forall x \in \overline{\Omega}$, $\forall k$) such that $h_k \rightharpoonup \gamma(u)$ in V (cf. [16, Proof of Theorem 2.1]). We now set $w = h_k$ in (3.9):

$$a(u, h_k) + \int_\Omega g_2(\cdot, u) h_k \, dx = 0 \quad \forall k,$$

and pass to the limit $k \to \infty$. By (3.8) and Lebesgue's theorem it results that

$$a(u, \gamma(u)) + \int_\Omega g_2(\cdot, u) \gamma(u) \, dx = 0. \tag{3.10}$$

By monotonicity of the function γ,

$$a(u, \gamma(u)) \ge 0. \tag{3.11}$$

Further we claim that

$$\int_\Omega [g_1(\cdot, u) - f] \gamma(u) \, dx \ge 0. \tag{3.12}$$

In fact, if $x \in \Omega$ is such that $R_1 \le u(x) \le R_2$, then $\gamma(u(x)) = 0$. For $x \in \Omega$ such that $u(x) < R_1$ we have

$$g_1(x, u(x)) = g(x, R_1) \leq f(x)$$

and

$$\gamma(u(x)) < 0.$$

Consequently the integrand in (3.12) is nonnegative there. The same is true on the set where $u(x) > R_2$. Inequality (3.12) thus follows.

In view of (3.11) and (3.12) we infer from (3.10) that

$$\int_\Omega \beta(u) \, \gamma(u) \, dx \leq 0.$$

Since $\beta(t) \gamma(t) \geq 0 \; \forall \; t \in \mathbb{R}$, with strict inequality whenever $t < R_1$ and $t > R_2$,

$$R_1 \leq u(x) \leq R_2 \quad \text{a.e. in } \Omega.$$

Hence $g(\cdot, u) = g_1(\cdot, u) - f$, and (3.8) and (3.9) say that u is a weak solution of problem (N).

4. PROOF OF PROPOSITION 3.2

The proof of Proposition 3.2 is based on the following lemma, which extends an earlier result by Strauss [17].

LEMMA 4.1. Let the function $g: \Omega \times \mathbb{R} \to \mathbb{R}$ satisfy conditions (BI, II), and let $\{v_n\}$ be a sequence in W such that $v_n \rightharpoonup u$ in V and $b(v_n, v_n) \leq \text{const} \; \forall \; n$. Then $u \in V_1$, $b(v_n, w) \to b(u, w)$ $\forall \; w \in W$, and $\int_\Omega g(\cdot, u) \, u \, dx \leq \liminf b(v_n, v_n)$.

NONLINEAR ELLIPTIC PROBLEMS

Proof. Let g be decomposed into a sum $g = g_0 + h$ of functions according to (3.5). By passing to a subsequence we may assume that $v_n \to u$ in $L^1(\Omega)$ as well as pointwise a.e. Since

$$\int_\Omega g_0(\cdot, v_n) v_n \, dx \leq \int_\Omega g(\cdot, v_n) v_n \, dx$$
$$+ \int_\Omega |h(\cdot, v_n)||v_n| \, dx \leq \text{const} \quad \forall \, n,$$

it follows by Fatou's lemma that

$$\int_\Omega g_0(\cdot, u) u \, dx \leq \liminf \int_\Omega g_0(\cdot, v_n) v_n \, dx \leq \text{const}.$$

Further, application of the Lebesgue theorem shows that

$$\int_\Omega h(\cdot, v_n) v_n \, dx \to \int_\Omega h(\cdot, u) u \, dx$$

and

$$\int_\Omega h(\cdot, v_n) w \, dx \to \int_\Omega h(\cdot, u) w \, dx \quad \forall \, w \in W.$$

We conclude that $g(\cdot, u) u \in L^1(\Omega)$. Since

$$|g_0(x, t)| = \frac{1}{|t|} g_0(x, t) t \quad (t \neq 0), \tag{4.1}$$

we estimate

$$|g_0(x, t)| \leq g_0(x, t) t + \sup_{|t| \leq 1} |g_0(x, t)| \quad (t \in \mathbb{R}).$$

Consequently $g_0(\cdot, u) \in L^1(\Omega)$, and thus $g(\cdot, u) \in L^1(\Omega)$, i.e.,

$$u \in V_1.$$

Again by (4.1), for given $\delta > 0$ and any function v_n, we have either

$$|v_n(x)| \leq \delta^{-1}$$

or

$$|g_0(x, v_n(x))| \leq \delta\, g_0(x, v_n(x))\, v_n(x).$$

Let $\Omega_0 \subset \Omega$ be any measurable subset. Then

$$\int_{\Omega_0} |g_0(x, v_n(x))|\, dx \leq \int_{\Omega_0} \sup_{|t| \leq \delta^{-1}} |g_0(x, t)|\, dx$$
$$+ \delta \int_{\Omega} g_0(x, v_n(x))\, v_n(x)\, dx.$$

It follows that the functions $g_0(\cdot, v_n)$ have equiabsolutely continuous integrals. Thus, by Vitali's convergence theorem (e.g., [18]),

$$g_0(\cdot, v_n) \to g_0(\cdot, u) \quad \text{in } L^1(\Omega),$$

and therefore

$$\int_{\Omega} g_0(\cdot, v_n)\, w\, dx \to \int_{\Omega} g_0(\cdot, u)\, w\, dx \quad \forall\, w \in W.$$

Putting together we obtain all the assertions of the lemma.

We now proceed to the following proof.

<u>Proof of Proposition 3.2.</u> Let $\{v_n\}$ in W be a sequence and $u \in V$, $g \in V^*$ elements such that

$$v_n \rightharpoonup u \quad \text{in } V,$$

$$Tv_n \rightharpoonup g|_W \quad \text{in } W^*,$$

$$\limsup (Tv_n, v_n) \leq (g, u).$$

Since

NONLINEAR ELLIPTIC PROBLEMS

$$\limsup (Bv_n, v_n) \le (g, u) + \limsup |(Av_n, v_n)| \le \text{const},$$

it follows by Lemma 4.1 that $u \in V_1 = \mathcal{D}(T)$ and $Bv_n \rightharpoonup Bu$ in W^*. Since the mapping $A: V \to V^*$ is bounded and V^* is reflexive, we may assume that $Av_n \rightharpoonup \chi$ in V^* (and thus $\tilde{A}v_n \rightharpoonup \chi|_W$ in W^*), with $\chi|_W + Bu = g|_W$. The quasi-boundedness of T follows.

In order to show that $\chi = Au$, it suffices by property (M) of A to prove that

$$\limsup (Av_n, v_n - u) \le 0. \qquad (4.2)$$

For any $h \in W$,

$$\limsup (Av_n, v_n - u)$$

$$= \limsup (Av_n, v_n - h) + (\chi, h - u)$$

$$= \limsup \bigl[(Tv_n, v_n) - b(v_n, v_n) - (Tv_n, h) + b(v_n, h)\bigr]$$
$$+ (\chi, h - u)$$

$$\le (g - \chi, u - h) - b(u, u) + b(u, h).$$

For each $M \ge 0$ let $u^M \in V$ be the function u truncated by $\pm M$, and let $\{h_k\} \subset C^\infty(\overline{\Omega})$ denote a sequence with $|h_k(x)| \le M$ ($\forall\, k$, $\forall\, x \in \overline{\Omega}$) such that $h_k \rightharpoonup u^M$ in V. Setting $h = h_k$ in (4.3) and passing to the limit $k \to \infty$ we obtain (since $u \in V_1$)

$$\limsup (Av_n, v_n - u) \le (g - \chi, u - u^M) - b(u, u) + b(u, u^M).$$

Finally let $M \to +\infty$ to get (4.2).

REFERENCES

1. F. E. Browder, Existence theorems for nonlinear partial differential equations, *Proc. Symp. Pure Math.*, Vol. 16, Amer. Math. Soc., Providence, Rhode Island, 1970.

2. J. P. Gossez, Nonlinear elliptic boundary value problems for equations with rapidly increasing coefficients, *Trans. Amer. Math. Soc.*, *190* (1974), 163-205.

3. F. E. Browder, Existence theory for boundary value problems for quasi-linear elliptic systems with strongly nonlinear lower order terms, *Proc. Symp. Pure Math.*, Vol. 23, Amer. Math. Soc., Providence, Rhode Island, 1973.

4. B. A. Ton, Pseudo-monotone operators in Banach spaces and nonlinear elliptic equations, *Math. Z.*, *121* (1971), 243-252.

5. P. Hess, On nonlinear mappings of monotone type with respect to two Banach spaces, *J. Math. Pures et Appl.*, *52* (1973), 13-26.

6. P. Hess, A strongly nonlinear elliptic boundary value problem, *J. Math. Anal. Appl.*, *43* (1973), 241-249.

7. P. Hess, Variational inequalities for strongly nonlinear elliptic operators, *J. Math. Pures et Appl.*, *52* (1973), 285-297.

8. P. Hess, On a class of strongly nonlinear elliptic variational inequalities, *Math. Ann.*, *211* (1974), 289-297.

9. B. A. Ton, On strongly nonlinear elliptic variational inequalities, *Pacific J. Math.*, *48* (1973), 279-291.

10. D. E. Edmunds, V. B. Moscatelli, and J. R. L. Webb, Strongly nonlinear elliptic operators in unbounded domains, *Publ. Math. Bordeaux*, *4* (1974), 6-32.

11. J. R. L. Webb, On the Dirichlet problem for strongly nonlinear elliptic operators in unbounded domains, *J. London Math. Soc.*, *10* (1975), 163-170.

12. M. A. Krasnoselskii, *Topological Methods in the Theory of Nonlinear Integral Equations* (Engl. Transl.), Pergamon Press, New York, 1964.

13. J. L. Lions, *Quelques méthodes de résolution des problèmes aux limites non linéaires*, Dunod, Gauthier-Villars, Paris, 1969.

14. P. Hess, On semi-coercive nonlinear problems, *Indiana Univ. Math. J.*, *23* (1974), 645-654.

15. M. Schatzman, Problèmes aux limites non linéaires, non coercifs, *Ann. Scuola Normale Superiore Pisa*, *23* (1973), 641-686.

16. S. Agmon, *Lectures on Elliptic Boundary Value Problem*, Van Nostrand Math. Studies, Van Nostrand, Princeton, 1965.

17. W. A. Strauss, On weak solutions of semi-linear hyperbolic equations, *Ann. Acad. Brasil. Cienc.*, 42 (1970), 645-651.
18. I. P. Natanson, *Theory of Functions of a Real Variable*. (Engl. Transl.), Ungar, New York, 1964.

UNFOLDING OF SINGULARITIES

Peter Hilton

Battelle Research Center
Seattle, Washington

and

Department of Mathematics
Case Western Reserve University
Cleveland, Ohio

1. INTRODUCTION

There are many reasons why mathematicians today should be making themselves aware of exciting new applications of mathematics. To list only two such reasons, mathematicians should be concerned to broaden their own competence, and they should be in a position to draw to their students' attention those parts of science where mathematics is leading to important new advances.

There can be little doubt that Thom's theory of *catastrophes*, as described in his great seminal work *Stabilité Structurelle et Morphogénèse*, is one of the most exciting advances in mathematical and scientific thought to have occurred for many years.

It is not surprising that it has attracted the attention not only of mathematicians but also of many scientists—indeed the Foreword to Thom's book is written by the eminent geneticist C. H. Waddington. However, it is certainly not only in the area of biology—and, more especially, morphogenesis—that applications are to be found of Thom's theory. Already applications are to be found in linguistics, economics, the study of heartbeat and of nerve impulses, fluid and gas dynamics, climatology, and even such apparently purely empirical fields of enquiry as the strategies of argumentation, women's liberation, and other conflict situations.

Thom's book refers frequently—one might even say tantalizingly—to many areas of application; indeed, the basic theme is quite as much philosophical as it is mathematical or scientific, since it treats of the very nature of the process whereby mathematics may be used to provide a model of the changing nature of the real world. In particular, the theory of catastrophes provides a framework for the discussion of those situations, among the most important in nature, in which a continuous change or perturbation of the control variables leads to a discontinuous change of outcome.

It is not surprising that Thom's book is difficult to read—particularly so for mathematicians. As John Guckenheimer has written in his very perceptive review of the book,[†] "René Thom has written a provocative book. It contains much of interest to mathematicians and has already had a significant impact upon mathematics, but it is not a work of mathematics." Thus there remains the practical problem of how mathematicians who are expert neither in the principal areas of mathematics utilized by Thom nor in the currently dominant fields of application (for example, the present author) can acquire a knowledge of the subject adequate for a comprehension, at least, of the fundamental mathematical ideas involved, and an ability to

---

[†] *Bull. Amer. Math. Soc.* (1973), 878-890.

convey to their students the importance of these ideas and an adequate preparation for their subsequent acquisition.

The present chapter represents a part of the author's attempt to acquire such knowledge. No claim whatsoever is made to originality, though the treatment does not coincide precisely with any with which the author is familiar. An attempt has been made to identify a key mathematical concept of Thom's theory— that of the universal unfolding of a singularity—and to indicate some of the important results associated with this concept, leading to a description of the seven elementary catastrophes of codimension ≤ 4. Even this modest objective would almost certainly have been beyond the author's reach had he not had the advantage of having available the notes by Theodor Bröcker (*Differenzierbare Abbildunge*, Universität Rogensburg) and of attending the lectures of Ioan James at the Summer Institute of the Australian Mathematical Society (Monash University, 1974). A further acknowledgment is, of course, due to René Thom himself. Not only is his book inspiring; he himself has been good enough to talk to the author several times (in particular, at Cornell University in 1969, and at Memorial University, St. John's, Newfoundland, in 1973) about his ideas and convey, at least, the enormous potential scope of his thought.

The author has a further reason to be grateful to the organizing committee of the São Paulo Symposium for their invitation to present this discussion. There is great pressure on mathematics departments all over the world to put more emphasis on the teaching of applied mathematics and the training of applied mathematicians. The author has no quarrel with this point of view, provided it is implemented intelligently. Unfortunately, there are those who appear to believe that there are certain areas of mathematics—for example, algebra, geometry, topology, number theory—which can be safely neglected (indeed, which *should* be neglected) by the student purposing to work in applied mathematics. This view, often stridently advocated by those specializing in fields belonging to the complement of the set described above, seems to the author wholly erroneous and highly prejudicial to the prospects for significant new advances in the applications of mathematics. The fact that an area of mathematics has not hitherto found application in the sciences (and here, of course, we include the physical, biological, and social sciences) by no means implies that this will remain the case; history, past and present, abounds with examples. The work of Thom provides, indeed, an incontrovertible refutation of the callow view referred to; for, even in the brief extract from that theory described in this chapter, there enter significantly important aspects of commutative algebra and differential topology. To prepare a student to work in Thom's theory, it is clear that one must train him (and oneself) to become a genuine mathematician, but one imbued with the desire to use mathematics to understand the nature of the real world. It would really appear that this last characteristic, much more closely related to motivation than to expertise, is what (if anything) should distinguish the training and the modus operandi of applied mathematicians from those of so-called pure mathematicians.

2. GERMS OF C^∞-FUNCTIONS

Consider the set S of functions $\mathbb{R}^n \to \mathbb{R}$ which are C^∞ in a neighborhood of the origin. We say that two such functions f, $g \in S$ determine the same *germ* if they agree in some neighborhood of the origin, so that a germ of a C^∞ function is, strictly, an equivalence class of C^∞ functions. Since our theory is entirely local, we will permit ourselves to speak of the *values* of a germ η and to write $\eta(x)$, $x \in \mathbb{R}^n$, although it would be more correct (though excessively pedantic) to choose a function from the equivalence class η.

DEFINITION 2.1. E_n is the ring of C^∞-germs $\eta: \mathbb{R}^n \to \mathbb{R}$; M_n is the ideal consisting of those germs η such that $\eta(0) = 0$. If $\eta \in M_n^2$ then η is *singular*; if $\eta \in M_n^3$ then η is *degenerate*.

UNFOLDING OF SINGULARITIES

PROPOSITION 2.2. The ideal M_n is generated by the coordinate functions. Moreover, M_n is the unique maximal ideal in E_n.

Proof. Let $\eta \in M_n$. Then

$$\eta(x) = \int_0^1 \frac{d}{dt}(\eta(xt))\,dt$$

$$= \int_0^1 \left(\sum_{i=1}^n x_i \frac{\partial}{\partial y_i} \eta(y)_{xt}\right) dt$$

$$= \sum_{i=1}^n \xi_i x_i,$$

where

$$\xi_i = \int_0^1 \left(\frac{\partial}{\partial y_i} \eta(y)_{xt}\right) dt.$$

Now if $\eta \notin M_n$, then there is a neighborhood of the origin in which $\eta(x) \neq 0$. Thus η^{-1} exists, and the only ideal of E_n containing η is E_n itself. This shows that M_n is the unique maximal ideal in E_n.

It follows from the first assertion of Proposition 2.2 that M_n^s is generated by the monomials in $x_1, x_2, \ldots, x_n$ of degree s. We will henceforth always assume that $\eta \in M_n^2$, that is, that η is a singular germ, although not all our definitions require this.

DEFINITION 2.3. The k-jet, $j^k(\eta)$, of η is the projection of η onto E_n/M_n^{k+1}. Thus $j^k(\eta)$ may be regarded as the Taylor expansion, up to degree k, of η.

DEFINITION 2.4. We say that η is *right-equivalent* to ξ, and write $\eta \sim \xi$, if η, ξ are C^∞-germs such that $\eta = \xi\varphi$ for some (local) diffeomorphism $\varphi: \mathbb{R}^n \to \mathbb{R}^n$ with $\varphi(0) = 0$.

DEFINITION 2.5. We say that η is *right-determined by its k-jet*, or, simply, *k-determined*, if $j^k(\eta) = j^k(\xi) \implies \eta \sim \xi$.

One can give a *geometric* definition of the codimension of the germ $\eta \in M_n^2$ as the codimension of the orbit of η under the group of diffeomorphisms φ entering into Definition 2.4. However, the following equivalent *algebraic* definition is usually more useful.

If $\eta \in M_n^2$, then its partial derivatives $\partial\eta/\partial x_i$, $i = 1, \ldots, n$, belong to M_n and we write $\langle \partial\eta/\partial x \rangle$ for the ideal generated by the $\partial\eta/\partial x_i$ and $\langle \partial\eta/\partial x \rangle \subseteq M_n$.

DEFINITION 2.6. The *codimension* of the singular germ η

$$\text{codim } \eta = \dim_R M_n / \langle \frac{\partial \eta}{\partial x} \rangle.$$

By abuse of language we will call a set of germs in M_n a *basis* for $M_n/\langle \partial\eta/\partial x \rangle$ if the projections of the elements of the set onto $M_n/\langle \partial\eta/\partial x \rangle$ constitute a basis.

Examples.

$n = 1$ $\eta = x^m$ codim $\eta = m - 2$
Basis $(x, x^2, \ldots, x^{m-2})$

$n = 2$ $\eta = x^2 y$ codim $\eta = \infty$

$n = 2$ $\eta = x^4 + y^4$ codim $\eta = 8$
Basis $(x, y, x^2, xy, y^2, x^2 y, xy^2, x^2 y^2)$

$n = 3$ $\eta = x^3 + y^3 + z^3$ codim $\eta = 7$
Basis $(x, y, z, xy, xz, yz, xyz)$.

We now state and prove one of the key theorems of the theory.

UNFOLDING OF SINGULARITIES 117

THEOREM 2.7 (Mather). If $M_n^k \subseteq M_n \langle \partial \eta / \partial x \rangle$, then the singular germ η is k-determined.

To prove this theorem we first need to invoke an algebraic lemma.

LEMMA 2.8 (Nakayama). Let R be a commutative unitary ring, A, B, C R-modules with A finitely generated and A, B $\subseteq$ C. Let M be an ideal of R such that $z \in M \implies 1 + z$ is invertible. Then

$$A \subseteq B + MA \implies A \subseteq B.$$

(Note that the role of C is simply to ensure that B + MA has meaning)

Proof of Lemma. First suppose B = 0. Let $a_1, a_2, \ldots, a_n$ generate A and let α be the vector $(a_1, a_2, \ldots, a_n)$. Then since $A \subseteq MA$ there is a matrix Z, with entries in M, such that $\alpha = Z\alpha$. Now $\det(1 - Z) = 1 + z$ with $z \in M$. Thus $1 - Z$ is invertible, so that $\alpha = 0$.

In the general case, the hypothesis $A \subseteq B + MA$ immediately yields $A/(A \cap B) \subseteq M \cdot A/(A \cap B)$. By what we have proved, $A/(A \cap B) = 0$ so that $A \subseteq B$.

Proof of Theorem. We are given that $M_n^k \subseteq M_n \langle \partial \eta / \partial x \rangle$ and $j^k(\eta) = j^k(\xi)$ and we wish to infer that $\eta \sim \xi$. Construct the homotopy $F_t(x) = F(x, t) = (1 - t) \eta(x) + t \xi(x)$, $0 \leq t \leq 1$. Then

$$\frac{\partial F}{\partial x_i} - \frac{\partial \eta}{\partial x_i} = t \left(\frac{\partial \xi}{\partial x_i} - \frac{\partial \eta}{\partial x_i} \right) \in M_n^k. \quad (2.1)$$

Now embed E_n, M_n in E_{n+1}, M_{n+1} in the obvious way. Then

$$M_n^k E_{n+1} \subseteq M_n \left\langle \frac{\partial \eta}{\partial x} \right\rangle E_{n+1} \subseteq M_n \left\langle \frac{\partial F}{\partial x} \right\rangle E_{n+1} + M_{n+1} M_n^k E_{n+1},$$

by (2.1). We apply the Nakayama Lemma with $R = E_{n+1}$, $M = M_{n+1}$, $A = M_n^k E_{n+1}$, $B = M_n \langle \partial F/\partial x \rangle E_{n+1}$, to ensure that

$$M_n^k E_{n+1} \subsetneq M_n \left\langle \frac{\partial F}{\partial x} \right\rangle E_{n+1}. \qquad (2.2)$$

Now $\partial F/\partial t = \xi - \eta \in M_n^{k+1}$; thus certainly $\partial F/\partial t \in M_n^k E_{n+1}$, so that we have $\xi_1, \ldots, \xi_n \in M_n E_{n+1}$, such that

$$\sum_{i=1}^{n} \frac{\partial F}{\partial x_i}(x, t) \xi_i(x, t) + \frac{\partial F}{\partial t}(x, t) = 0. \qquad (2.3)$$

The differential equation

$$\frac{\partial H}{\partial t}(x, t) = \xi(H(x, t), t), \quad H: \mathbb{R}^n \times \mathbb{R} \to \mathbb{R}^n, \qquad (2.4)$$

has a unique solution near $t = t_0$, say, subject to $H_{t_0} = \text{Id}$. Note, moreover, that $H(0, t) = 0$ since $\xi(0, t) = 0$. From (2.3) and (2.4) it follows that

$$\frac{d}{dt} F(H(x, t), t) = 0,$$

so that $F_t H_t$ is independent of t near t_0. Thus

$$F_t H_t = F_{t_0};$$

moreover, H_t is a diffeomorphism, since $H_{t_0} = \text{Id}$. Thus $F_t \sim F_{t_0}$ for t sufficiently close to t_0. The usual argument, invoking the compactness of the unit interval, now shows that $F_0 \sim F_1$, or $\eta \sim \xi$.

Remark. It may be seen that a germ is k-determined if its k-jet is k-determined. This will be a useful fact when we come to catalog the universal unfoldings of codimension ≤ 4. It may also be shown that every germ of finite codimension is k-determined for some k.

UNFOLDING OF SINGULARITIES

THEOREM 2.9 (Splitting Lemma). Let $\eta \in M_n^2$ with codim $\eta < \infty$. Then

$$\eta(x_1, x_2, \ldots, x_n) \sim \xi(x_1, x_2, \ldots, x_r) + q(x_{r+1}, \ldots, x_n),$$

where $\xi \in M_r^3$ and q is a nondegenerate quadratic form.

We will not prove this theorem here, but will make some comments on its proof in the next section. We note however, that, since q may, by a change of coordinates, be put in the form $\pm x_{r+1}^2 \pm \cdots \pm x_n^2$, it follows from Definition 2.6 that

$$\text{codim } \eta = \text{codim } \xi. \qquad (2.5)$$

In view of the Splitting Lemma, the following elementary result is highly relevant to our purposes.

PROPOSITION 2.10. If $\eta \in M_n^3$ then codim $\eta \geq \binom{n+1}{2}$.

Proof. Now dim $j^2 M_n = n + \binom{n+1}{2}$. However, since $\eta \in M_n^3$, it is plain that dim $j^2 \langle \partial \eta / \partial x \rangle \leq n$. Thus

$$\text{codim } \eta = \dim M_n / \langle \tfrac{\partial \eta}{\partial x} \rangle \geq \dim j^2 M_n / j^2 \langle \tfrac{\partial \eta}{\partial x} \rangle \geq \binom{n+1}{2}.$$

We will use this proposition to justify the restriction $n \leq 2$ if we wish to confine attention to degenerate germs of codimension ≤ 5.

3. UNFOLDINGS OF A SINGULARITY

We fix a germ $\eta \in M_n$. Then a germ $f \in M_{n+r}$ is called an *unfolding* of η if

$$f(x, 0) = \eta(x), \quad x \in \mathbb{R}^n.$$

We may write $f_u(x) = f(x, u)$, $x \in \mathbb{R}^n$, $u \in \mathbb{R}^r$, and think of the unfolding f as an r-parameter family of germs. We call r the *codimension* of the unfolding f; we may write (f, r) for f to emphasize the codimension.

Examples.

1. The *constant* unfolding of codimension r is given by $f(x, u) = \eta(x)$, $x \in \mathbb{R}^n$, $u \in \mathbb{R}^r$.

2. Let $b_1, \ldots, b_r \in M_n$. Then we obtain an unfolding f of codimension r, the sum of one-parameter unfoldings, by the formula

$$f(x, u) = \eta(x) + \sum_{j=1}^{r} b_j(x) u_j. \tag{3.1}$$

Our second example illustrates implicitly the notion of the sum of two unfoldings of η; thus we define

$$(f, r) + (g, s) = (h, r + s),$$

where

$$h(x, u, v) = f(x, u) + g(x, v) - \eta(x). \tag{3.2}$$

DEFINITION 3.1. Let (f, r), (g, s) be unfoldings of η. A *morphism* $(\varphi, \alpha): (f, r) \to (g, s)$ consists of the following:

UNFOLDING OF SINGULARITIES

(a) A germ $\varphi: \mathbb{R}^{n+s} \to \mathbb{R}^{n+r}$, such that $\varphi(x, 0) = (x, 0)$, which is *fibrewise* in the sense that there is a germ $\bar{\varphi}: \mathbb{R}^s \to \mathbb{R}^r$ giving rise to a commutative diagrama

$$\begin{array}{ccc} \mathbb{R}^{n+s} & \overset{\varphi}{\to} & \mathbb{R}^{n+r} \\ \downarrow \pi & & \downarrow \pi \\ \mathbb{R}^s & \overset{\bar{\varphi}}{\to} & \mathbb{R}^r \end{array}$$

π standing for the obvious projections; and

(b) a germ $\alpha \in M_s$ such that $g = f\varphi + \alpha\pi$.

Notice that the condition that φ be fibrewise simply says that if $\varphi(x, v) = (y, u)$, then u depends only on v. Notice also that we may restate (b) in terms of the s-parameter family of translations $a_v(t) = t + \alpha(v)$. Then

$$g_v = a_v f \varphi_v. \qquad (3.3)$$

It would be possible to replace these translations by more general transformations, but we follow Mather in confining ourselves to translations in this definition.

Now, given a morphism $(\varphi, \alpha): (f, r) \to (g, s)$ and a morphism $(\psi, \beta): (g, s) \to (h, t)$, we may compose morphisms by the rule

$$(\psi, \beta)(\varphi, \alpha) = (\varphi\psi, \alpha\bar{\psi} + \beta), \qquad (3.4)$$

and in this way we obtain a *category* of unfoldings of η. This leads immediately to a notion of *equivalence* of unfoldings.

However, we may also take the point of view that if $(\varphi, \alpha): (f, r) \to (g, s)$, then (g, s) is *induced* from (f, r) by means of (φ, α). For it is obvious from Definition 3.1 that (f, r) and (φ, α) determine (g, s). This point of view leads to the following important definition.

DEFINITION 3.2. An unfolding f of η is *versal* if it induces every other unfolding of η. A versal unfolding of minimum codimension is *universal*.

Then a key theorem in the theory is the following.

THEOREM 3.3. Let η be a singular germ of finite codimension. Then η admits a versal unfolding.

The method of proof of this theorem, and the means of finding universal foldings, are by way of the geometrical concept of k-*transversal* unfoldings. An unfolding (f, r) of η is k-transversal if the k-jet (with respect to the x variable), $j_1^k f$, is transversal at 0 to the orbit of η under right transformations (see Definition 2.4). However, we will adopt an algebraic characterization of k-transversality which will enable us to show more easily the connection with versality. We assume henceforth that the germ η has finite codimension.

Let V_f be the subspace of E_n spanned by the germs

$$\frac{\partial f}{\partial u_j} \mid \mathbb{R}^n \times 0 - \frac{\partial f}{\partial u_j}(0, 0), \quad j = 1, 2, \ldots, r.$$

DEFINITION 3.4. The unfolding (f, r) of η is k-*transversal* if

$$M_n \subseteq \langle \frac{\partial \eta}{\partial x} \rangle + V_f + M_n^{k+1}.$$

Then the following propositions are evident.

PROPOSITION 3.5. Let $b_1, \ldots, b_r$ be elements of M_n which project to a basis for $M_n / \langle \partial \eta / \partial x \rangle + M_n^{k+1}$. Then the unfolding of Example (2) is k-transversal.

UNFOLDING OF SINGULARITIES

PROPOSITION 3.6. The sum of any unfolding and a k-transversal unfolding is k-transversal.

We will prove

THEOREM 3.7. A versal unfolding of a singular germ η of finite codimension is k-transversal for all k.

Proof. Let (f, r) be versal. Since η has finite codimension, there certainly is a k-transversal unfolding (g, s), and hence a morphism

$$(\varphi, \alpha): (f, r) \to (g, s), \quad g = f + \alpha\pi.$$

It is plain that $V_g = V_{f\varphi}$; we will show that

$$V_{f\varphi} \subseteq \left\langle \frac{\partial \eta}{\partial x} \right\rangle + V_f. \qquad (3.5)$$

Then

$$M_n \subseteq \left\langle \frac{\partial \eta}{\partial x} \right\rangle + V_{f\varphi} + M_n^{k+1} \subseteq \left\langle \frac{\partial \eta}{\partial x} \right\rangle + V_f + M_n^{k+1},$$

and so (f, r) is k-transversal.

Thus it remains to prove (3.5). Suppose $\varphi: (x, v) \to (y, u)$, $x, y \in \mathbb{R}^n$, $u \in \mathbb{R}^r$, $v \in \mathbb{R}^s$. Then, since $\varphi(x, 0) = (x, 0)$,

$$\frac{\partial f\varphi}{\partial v_j}(x, 0) =$$

$$\sum_i \frac{\partial f}{\partial x_i}(x, 0) \frac{\partial y_i}{\partial v_j}(x, 0) \qquad (3.6)$$

$$+ \sum_k \frac{\partial f}{\partial u_k}(x, 0) \frac{\partial u_k}{\partial v_j}(x, 0).$$

Now we note (i) $(\partial f/\partial x_i)(x, 0) = \partial \eta/\partial x_i$, (ii) u_k does not depend on x. Thus, recalling that η is singular,

$$\frac{\partial f\varphi}{\partial v_j} \mid \mathbb{R}^n \times 0 - \frac{\partial f\varphi}{\partial v_j}(0, 0)$$

$$= \sum_i \frac{\partial \eta}{\partial x_i} \frac{\partial y_i}{\partial v_j} (x, 0) \qquad (3.7)$$

$$+ \sum_k \left(\frac{\partial f}{\partial u_k} \mid \mathbb{R}^n \times 0 - \frac{\partial f}{\partial u_k} (0, 0) \right) \frac{\partial u_k}{\partial v_j} (0).$$

Plainly (3.5) follows from (3.7).

We state without proof the following fundamental theorem, leading to a converse of Theorem 3.7.

THEOREM 3.8. If η is k-determined, then any two k-transversal unfoldings of η of the same codimension are equivalent.

THEOREM 3.9. If η is k-determined, then every k-transversal unfolding of η is versal.

Proof. It is easy to see that, in the category of unfoldings of any germ η, there are always morphisms

$$f + g \to g,$$
$$f \to f + \text{constant}.$$

Now let f be k-transversal and let g be an arbitrary unfolding of η. Then we have an equivalence of unfoldings

$$f + \text{constant} \approx f + g,$$

since each is k-transversal by Proposition 3.6. Thus we have

$$f \to f + \text{constant} \approx f + g \to g,$$

showing that f is versal.

Similarly, we prove

THEOREM 3.10. Any versal unfolding of η is equivalent to the sum of a universal unfolding and a constant unfolding.

From the results above, we readily deduce the result which forms the basis of our arguments in the next section.

THEOREM 3.11. Let $\eta \in M_n^2$ have codimension r and let b_1, b_2, ..., b_r be a basis for $M_n/\langle\partial\eta/\partial x\rangle$. Then the unfolding (3.1), which is of codimension r, is universal.

Our final observation in this section is designed to indicate the line of proof of the Splitting Lemma.

PROPOSITION 3.12. If $\eta(x_1, x_2, \ldots, x_n) = \xi(x_1, x_2, \ldots, x_r) + q(x_{r+1}, \ldots, x_n)$, where $\eta \in M_n^2$, $\xi \in M_r^2$, and q is a nondegenerate quadratic form, and if f is a universal unfolding of ξ, then $f + q$ is a universal unfolding of η.

Proof. It is easy to see, by reducing q to a sum of squares that $M_n/\langle\frac{\partial\eta}{\partial x}\rangle$ and $M_r/\langle\frac{\partial\xi}{\partial x}|_r\rangle$ have the same basis; here we write $\langle\frac{\partial\xi}{\partial x}|_r\rangle$ for the subideal of M_r generated by $\partial\xi/\partial x_1$, ..., $\partial\xi/\partial x_r$. The proposition now follows from Theorem 3.11.

The Splitting Lemma, together with Proposition 3.12, shows that we may concentrate on degenerate germs in computing universal unfoldings.

4. THE SEVEN ELEMENTARY CATASTROPHES FOR SINGULARITIES OF CODIMENSION ≤ 4

Let $\eta \in M_n^2$ and let f be an unfolding of η of codimension r. We define its *set of singularities*, $\Sigma(f)$, by

$$\Sigma(f) = \{(x, u) \in \mathbb{R}^n \times \mathbb{R}^r \mid d_x f(x, u) = 0\}. \qquad (4.1)$$

We further define the *set of degenerate singularities* of f, $\Delta(f)$, by

$$\Delta(f) = \{(x, u) \in \mathbb{R}^n \times \mathbb{R}^r \mid d_x f(x, u) = 0, d_x^2 f(x, u) = 0\}. \qquad (4.2)$$

Finally, we define the *catastrophe set*, of f to be the projection of $\Delta(f)$ onto $\mathbb{R}^r$. In the light of the Splitting Lemma we will feel free to concentrate on the case $\eta \in M_n^3$ so that $0 \in D(f)$ for all unfoldings of η. We will be looking at universal unfoldings which provide us with "minimal" or *elementary* catastrophe sets for unfoldings of η. Also our universal unfoldings shall be of codimension ≤ 4, so that η itself will, by Theorem 3.11, have codimension ≤ 4, and hence, by Proposition 2.10, $n \leq 2$.

First suppose $n = 1$. Then $\eta \sim \pm x^m$, codim $\eta = m - 2$, so that $3 \leq m \leq 6$, since we require codim $\eta \leq 4$. These possibilities give rise to the following universal unfoldings (we take $\eta = x^m$):

$$\begin{aligned}
m = 3 \quad & f(x, u) = x^3 + ux \\
& \text{(fold catastrophe)} \\
m = 4 \quad & f(x, u) = x^4 + u_1 x^2 + u_2 x \\
& \text{(cusp catastrophe)} \\
m = 5 \quad & f(x, u) = x^5 + u_1 x^3 + u_2 x^2 + u_3 x \\
& \text{(swallowtail catastrophe)} \\
m = 6 \quad & f(x, u) = x^6 + u_1 x^4 + u_2 x^3 + u_3 x^2 + u_4 x \\
& \text{(butterfly catastrophe)}.
\end{aligned} \qquad (4.3)$$

UNFOLDING OF SINGULARITIES

Let us examine more closely the cusp catastrophe, also known as the Riemann-Hugoniot catastrophe, first studied in connection with the behavior of gas under pressure (see Fig. 1). In this case, we have

$$\Sigma(f) = \{(x, u) \mid 4x^3 + 2u_1 x + u_2 = 0\},$$

$$\Delta(f) = \{(x, u) \mid 4x^3 + 2u_1 x + u_2 = 0, 6x^2 + u_1 = 0\},$$

$$D(f) = \{(u_1, u_2) \mid 8u_1^3 + 27u_2^2 = 0\}.$$

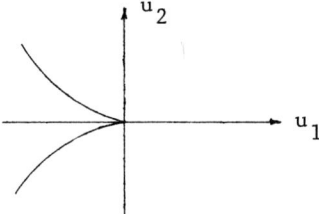

FIG. 1. $8u_1^3 + 27u_2^2 = 0$; the cusp catastrophe.

Inside the cusp, $8u_1^3 + 27u_2^2 < 0$ and the equation $4x^3 + 2u_1 x + u_2 = 0$ has three real roots; outside the cusp, $8u_1^3 + 27u_2^2 > 0$ and the equation $4x^3 + 2u_1 x + u_2 = 0$ has one real root. This shows how the behavior of the singular set of the function $x^4 + u_1 x^2 + u_2 x$ depends on the parameters u_1, u_2; and, in particular, how it changes as we cross the catastrophe set. We may get a geometric picture of this behavior by drawing the "surface" $4x^3 + 2u_1 x + u_2 = 0$ (see Fig. 2).

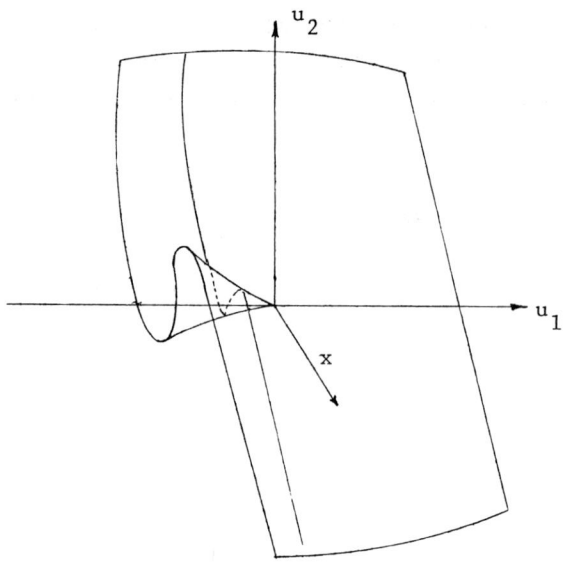

FIG. 2. $4x^3 + 2u_1 x + u_2 = 0$; the singular set of the universal unfolding.

We turn now to the case $n = 2$. Thus $\eta \in M_2^3$ and codim $\eta = 3$ or 4. We base our argument on an analysis of $j^3(\eta)$, the 3 jet of η. Thus $j^3(\eta)$ is a homogeneous polynomial in x, y of degree 3, and we may factorize over C.

$$j^3(\eta) = (a_1 x + b_1 y)(a_2 x + b_2 y)(a_3 x + b_3 y), \quad a_i, b_i \in C. \quad (4.4)$$

We distinguish five cases:

<u>Case 1</u>. *The factors in (4.4) are real, pairwise independent.*

In this case it is plain that $j^3(\eta)$ is right-equivalent to $x(x^2 - y^2)$. Moreover,

UNFOLDING OF SINGULARITIES

$$(x, y)^3 \subseteq (x, y)(3x^2 - y^2, xy),$$

so that, by Theorem 2.7, $j^3(\eta)$ is 3-determined; hence η is itself 3-determined. Applying Theorem 3.11, we see that we obtain the universal unfolding[†] of $j^3(\eta)$, of codimension 3,

$$f(x, y, u) = x(x^2 - y^2) + u_1(x^2 + y^2) + u_2 x + u_3 y$$

$$\text{(elliptic umbilic)}.$$

Case 2. *The factors in (4.4) are distinct, with one factor real.*

In this case it is plain that $j^3(\eta)$ is right-equivalent to $x^3 + y^3$. Moreover,

$$(x, y)^3 \subseteq (x, y)(x^2, y^2),$$

so that, again by Theorem 2.7, $j^3(\eta)$ is 3-determined, η is 3-determined, and the universal unfolding of $j^3(\eta)$, of codimension 3, is

$$f(x, y, u) = x^3 + y^3 + u_1 xy + u_2 x + u_3 y \quad \text{(hyperbolic umbilic)}.$$

Case 3. *The factors in (4.4) are real and two coincide.*

Now $j^3(\eta)$ is right-equivalent to $x^2 y$, which is *not* of finite codimension. Since η is of finite codimension, there must exist $k \geq 4$ such that $j^{k-1}(\eta) \sim x^2 y$, $j^k(\eta) \sim x^2 y + h(x, y)$,

---

[†] We do not distinguish between equivalent germs in describing the universal unfoldings.

where h is a homogeneous polynomial of degree k. Make the transformation $x \to x + \varphi$, $y \to y + \psi$, where φ, ψ are homogeneous polynomials of degree $k - 2 \geq 2$. This transformation is thus an origin-preserving diffeomorphism near 0, so that

$$j^k(\eta) \sim (x + \varphi)^2(y + \psi) + h(x + \varphi, y + \psi). \qquad (4.5)$$

The k-jet of the right-hand side of (4.5) is $x^2y + 2xy\varphi + x^2\psi + h(x, y)$. If

$$h(x, y) = \sum_{i=0}^{k} a_i x^{k-i} y^i,$$

we take

$$\varphi = -\frac{1}{2} \sum_{i=1}^{k-1} a_i x^{k-1-i} y^{i-1}, \quad \psi = -a_0 x^{k-2},$$

so that

$$j^k(\eta) \sim x^2 y + a_k y^k, \quad a_k \neq 0.$$

Now $x^2 y \pm y^k$ is k-determined and the codimension of $x^2 y \pm y^k$ is k [a suitable basis for $M_2/(xy, x^2 + y^{k-1})$ is provided by $(x, y, y^2, \ldots, y^{k-1})$]. Since we require the codimension to be ≤ 4, we must have $k = 4$. Thus, in this case, η is 4-determined, codim $\eta = 4$, and the universal unfolding of $j^4(\eta)$ is

$$f(x, y, u) = x^2 y + y^4 + u_1 y^3 + u_2 y^2 + u_3 x + u_4 y$$

(parabolic umbilic).

Case 4. *The factors in (4.4) all coincide.*

In this case $j^3(\eta) \sim x^3$, which is not of finite codimension. Then $j^4(\eta) \sim x^3 + h(x, y)$, where h is homogeneous of degree 4 (perhaps $h = 0$). We will show that this forces

UNFOLDING OF SINGULARITIES   131

codim $\eta \geq 5$, so that the codimension of the universal unfolding is outside the agreed range.

For dim $j^3 M_2 = 9$, while dim $j^3 \langle \partial \eta / \partial x \rangle \leq 4$. This last inequality arises because $j^3 \langle \partial \eta / \partial x \rangle$ is generated by $3x^2 + \partial h / \partial x$, $\partial h / \partial y$, x^3, $x^2 y$. It follows that

$$\text{codim } \eta = \dim_R M_2 / \langle \tfrac{\partial \eta}{\partial x} \rangle \geq \dim_R j^3 M_2 / \langle \tfrac{\partial \eta}{\partial x} \rangle \geq 5.$$

<u>Case 5</u>. $j^3(\eta) = 0$.

Then $j^4(\eta) = h(x, y)$, where h is homogeneous of degree 4. A similar argument to that above shows that in this case codim $\eta \geq 7$.

Thus we have obtained a complete list of the (seven) elementary catastrophes corresponding to universal unfoldings of codimension ≤ 4.

5. ACKNOWLEDGMENTS

The author has benefited very much from seeing the lecture notes "Differenzierbare Abbildungen" of Theodor Bröcker, being the notes of a course given in the summer of 1972 at Freiburg im Breisgau. He is also very grateful to Ioan James, who provided the original distillation of a portion of Bröcker's notes (in lectures given at the Summer Research Institute of the Australian Mathematical Society). The insights provided by James's lectures have been invaluable in preparing this brief summary.

BIBLIOGRAPHY

Bröcker, in the lecture notes referred to above, provides an excellent bibliography, which we reproduce here. To that bibliography should, of course, be added Bröcker's lecture notes themselves, available from the Fachbereich Mathematik, Universität Regensburg, Regensburg, West Germany. We have also added item 47, which appeared after Bröcker's notes, and which contains an excellent account of the applications of catastrophe theory (together with some beautiful pictures!).

By René Thom

1. *Stabilité Structurelle et Morphogenèse*, Benjamin, New York, 1972.
2. Une théorie dynamique de la morphogenèse, *T.T.B.*[†], *1*, 152-166.
3. With C. H. Waddington, Correspondence, *T.T.B.*, *1*, 166-179.
4. Topologie et signification, *L'Age de Science*, 4 (1968), 219-242.
5. Topological models in biology, *T.T.B.*, *3*, 89-116; *Topology*, 8 (1969), 313-336.
6. A mathematical approach to morphogenesis: archetypal morphologies, *Wistar Institute Symposium Monograph No. 9* (1969), 165-174.
7. Topologie et linguistique, *Essays on Topology and Related Subjects*, Springer-Verlag, New York, 1970, 226-248.
8. Structuralism and biology, *T.T.B.*, *4*, 68-82.
9. Modèles Mathématiques de la morphogenèse, Chapters 1-3, mimeographed, I.H.E.S., 1970-1971.
10. Les Symétries brisées en physique macroscopique et la mécanique quantique (to appear).
11. A *Dynamical Scheme for Vertebrate Embryology*, A.A.A.S., Philadelphia, 1971.
12. Sur les équations différentielles multiformes et leurs intégrales singulières (to appear).

[†] Note: T.T.B. = *Towards a Theoretical Biology* (C. H. Waddington, ed.), Edinburgh, 1970-1972.

13. Sur la typologie des langues naturelles: essai d'identification psycholinguistique, in *Formal analysis of natural Languages* (Menton, ed.), Proc. I.R.I.A. Congress, 1970; *Janua Linguarium* (to appear).

14. Language et catastrophes, in *Salvador Symposium on Dynamical Systems* (M. Peixoto, ed.), Academic Press, New York, 1973.

15. La théorie des catastrophes: état présent et perspectives (to appear).

By E. C. Zeeman

16. Geometry of catastrophe, *Times Literary Supplement*, 10, Dec. 1971.

17. Differential equations for the heartbeat and nerve impulse, *T.T.B.*, 4, 8-67.

18. A catastrophe machine, *T.T.B.*, 4, 276-282.

19. Catastrophe theory in brain modelling, *Conference on Neural Networks*, I.C.T.P., Trieste, 1972 (to appear in *J. Neuroscience*).

20. With C. Isnard, Some models from catastrophe theory, *Conf. Models in Social Sciences*, Edinburgh, 1972 (to appear).

21. With P. J. Harrison, Applications of catastrophe theory to macroeconomics, *Symp. Applications of Global Analysis*, Utrecht, 1973 (to appear).

22. On the unstable behaviour of stock exchanges, mimeographed, Warwick (1973) (to appear in *J. Math. Econ.*).

23. Applications of catastrophe theory, preprint, 1973.

Miscellaneous

24. R. Abraham, Introduction to Morphology, *Quatrième Rencontre entre Mathématiciens et Physiciens* (1972), Vol. 4, Fasc. 1, Publ. du Département de mathématiques de l'Université de Lyon, Tome 9 (1972), Fasc. Suppl. 1, 38-114.

25. V. I. Arnold, On braids of algebraic functions and the cohomology of "swallowtails," *Uspehi Mat. Nauk*, 23, 4 (1968), 247-248.

26. V. I. Arnold, On matrices depending on parameters, *Russian Math. Surveys*, 26, 2 (1971), 29-43 [transl. from *Uspehi Mat. Nauk*, 26, 2 (1971), 101-114].

27. V. I. Arnold, Lectures on bifurcations and versal families, *Russian Math. Surveys*, 27, 5 (1972) [transl. from *Uspehi Mat. Nauk*, 27, 5 (1972), 119-184].

28. V. I. Arnold, Integrals of rapidly oscillating functions and singularities of projections of Lagrangian manifolds, *Functional Anal. Appl.*, 6 (1973), 222-224 [transl. from *Funkcional. Anal. i Priložen.*, 6, 3 (1972), 61-62].

29. V. I. Arnold, Normal forms for functions near degenerate critical points, the Weyl Groups of A_k, D_k, E_k, and Lagrangian singularities, *Functional Anal. Appl.*, 6 (1973), 254-272 [transl. from *Funkcional. Anal. i Priložen.*, 6, 4 (1972), 3-25].

30. V. I. Arnold, Classification of unimodal critical points of functions, *Funkcional. Anal. i Priložen.*, 7, 3 (1973), 75-76.

31. N. A. Baas, On the models of Thom in Biology and Morphogenesis, lecture notes, Univ. Virginia,

32. D. H. Fowler, The Riemann-Hugoniot catastrophe and van der Waal's equation, *T.T.B.*, 4, 1-7.

33. A. M. Gabrielov, Intersection matrices for certain singularities, *Funkcional. Anal. i Priložen.*, 7, 3 (1973), 18-32.

34. A. N. Godwin, Three-dimensional pictures for Thom's parabolic umbilic, *Inst. Hautes Etudes Sci. Publ. Math.*, 40 (1971), 117-138.

35. J. Guckenheimer, Bifurcation and catastrophe (to appear).

36. J. Guckenheimer, Catastrophes and partial differential equations (to appear).

37. J. Mather, Right equivalence, manuscript.

38. F. Pham, Remarque sur l'equisingularité universelle, mimeographed, University of Nice, 1970.

39. F. Pham, Classification des Singularités, *Douzième Rencontre entre Mathématiciens et Physiciens* (1971). R.C.P. No. 25, Strasbourg.

40. T. Poston and A. E. R. Woodcock, On Zeeman's catastrophe machine, *Proc. Camb. Phil. Soc.*, 74 (1973), 217-226.

41. L. S. Shulman and M. Revzen, Phase transitions as catastrophes, *Collective Phenomena*, 1 (1972), 43-47.

42. D. Siersma, Singularities of C^∞ functions of right-codimension smaller than or equal to eight, *Indag. Math.*, 25 (1973), 31-37.

43. F. Takens, Singularities of functions and vector fields, *Nieuw Arch. Wisk.* (3), 20 (1972), 107-130.

44. G. N. Tyurina, Resolution of singularities of plane deformations of double rational points, *Functional Anal. Appl.*, 4 (1970), 68-73 [transl. from *Funkcional Anal. i Priložen.* 4, 1 (1970), 77-83].

45. G. Wassermann, Stability of unfoldings, *Lecture Notes in Mathematics*, Vol. 393, Springer-Verlag (1974).

46. G. Wassermann, (r, s)-stability of unfoldings, in preparation.

47. A. E. R. Woodcock and T. Poston, A geometrical study of the elementary catastrophes, *Lecture Notes in Mathematics*, Vol. 373, Springer-Verlag (1974).

THE DIRICHLET AND SUBSTITUTION FORMULAS FOR RIEMANN-STIELTJES INTEGRALS IN BANACH SPACES

Chaim Samuel Hönig

Instituto de Matemática e Estatística
Universidade de São Paulo
São Paulo, Brazil

An equality of the form

$$\int_a^b \left[\int_a^s d\alpha(t) \cdot h(t, s)\right] d\beta(s) = \int_a^b d\alpha(t) \left[\int_t^b h(t, s) \, d\beta(s)\right]$$

for iterated Riemann-Stieltjes integrals is called a *Dirichlet formula*.

An equality of the form

$$\int_a^b d\left[\int_a^t d\alpha(s) \cdot h(s)\right] g(t) = \int_a^b d\alpha(t) \cdot h(t) \, g(t)$$

for Riemann-Stieltjes integrals is called a *substitution formula*.

Without some restriction these formulas are not valid. In [4], p. 13 for instance, we give an example, with $\alpha(s) \equiv s$, where the formula of substitution is not true. A counterexample for the formula of Dirichlet is given in the Remark preceding Theorem 2.1.

In Secs. 2, 3, and 5 of this chapter we prove these formulas under adequate hypothesis and for different types of integrals (in Banach spaces). We need these formulas in our study of differential and integral equations with linear constraints (that is, generalized boundary conditions); see [1-4]. The different types of Riemann-Stieltjes integrals we use are necessary for the integral representation formulas of certain classes of linear operators (see [1, Theorems 2, 3, and 4], [3, Theorem 1], and [4]); however neither the representing function α nor the function h or g is necessarily measurable (in the sense of Bochner and Lebesgue), and this fact makes their study more difficult. For instance, the identical automorphism of $C([a, b])$ has the representation

$$\varphi = \int_a^b \varphi(t) \, d\alpha(t), \quad \varphi \in C([a, b]),$$

where $\alpha(t) = \chi_{]a,t]}$, $t \in [a, b]$, and the function

$$\alpha: t \in [a, b] \longmapsto \chi_{]a,t]} \in L_\infty([a, b])$$

is not measurable (see [2, p. 59]).

In Sec. 4 we take α of bounded variation (hence measurable) and consider a more restricted class of functions h for which the analog of the Darboux criterion for integrability is valid (and which are measurable).

Sections 2 to 5 are independent of each other and may be read in any order; only Sec. 5 uses some results of Sec. 4.

1. BASIC DEFINITIONS

In order to make this discussion more self-sufficient we repeat here many of the definitions given in [4] (see also [3, 1, 2]).

DIRICHLET AND SUBSTITUTION FORMULAS

DEFINITION 1.1. Given an interval $[a, b]$ of the real line, a *division* d of $[a, b]$ is a finite sequence $t_0 = a < t_1 < t_2 < \ldots < t_n = b$; we write $|d| = n$ and $\Delta d = \sup\{|t_i - t_{i-1}| \mid i = 1, 2, \ldots, |d|\}$. D or $D_{[a,b]}$ denotes the set of all divisions of $[a, b]$.

For every $\varepsilon > 0$ we define $D_\varepsilon = \{d \in D \mid \Delta d < \varepsilon\}$; the class $\{D_\varepsilon \mid \varepsilon > 0\}$ is a filter basis on D.

We say that a division d_2 is *finer* than a division d_1, and we write $d_1 \leq d_2$, if every point of d_1 is in d_2. For $d \in D$ we define $D_d = \{d' \in D \mid d \leq d'\}$; the class $\{D_d \mid d \in D\}$ is a filter basis on D finer than the preceding one and therefore we have

PROPOSITION 1.1. Given a topological space X and a function $f: D \to X$, the existence of $\lim_{\Delta d \to 0} f(d)$, i.e., according to the first filter basis, implies the existence of $\lim_{d \in D} f(d)$, i.e., according to the second filter basis, and we have the equality of both limits.

Given a Banach space X and a function $f: [a, b] \to X$, for every $d \in D$ we define the *oscillations*

$$\omega_i(f) = \sup\{\|f(t) - f(s)\| \mid t, s \in [t_{i-1}, t_i]\}$$

$$i = 1, 2, \ldots, |d|$$

and the *interior oscillations*

$$\omega_i^\bullet(f) = \sup\{\|f(t) - f(s)\| \mid t, s \in]t_{i-1}, t_i[\}$$

$$i = 1, 2, \ldots, |d|.$$

We write $\omega_d(f) = \sup\{\omega_i(f) \mid i = 1, 2, \ldots, |d|\}$ and $\omega_d^\bullet(f) = \sup\{\omega_i^\bullet(f) \mid i = 1, 2, \ldots, |d|\}$. Given $\delta > 0$ we define

$$\omega_\delta(f) = \sup\{\|f(t) - f(s)\| \mid t, s \in [a, b], |t - s| \leq \delta\}.$$

We denote by $\mathcal{D}$ the set of all pairs (d, ξ), where $d \in D$ and $\xi = (\xi_1, \ldots, \xi_{|d|})$ with $\xi_i \in [t_{i-1}, t_i]$. $\mathcal{D}^\bullet$ denotes the set of all pairs $(d, \xi^\bullet)$, where $d \in D$ and $\xi^\bullet = (\xi_1^\bullet, \ldots, \xi_{|d|}^\bullet)$ with $\xi_i^\bullet \in \,]t_{i-1}, t_i[\,$.

If t and s are real numbers, we write $t \wedge s = \inf(t, s)$ and $t \vee s = \sup(t, s)$. Given real numbers $c \leq d$, $|c, d|$ denotes any of the intervals $[c, d]$, $[c, d[$, $]c, d]$ and $]c, d[$. If A is a subset of $\mathbb{R}$, χ_A denotes its characteristic function: $\chi_A(t) = 1$ if $t \in A$ and $\chi_A(t) = 0$ if $t \notin A$.

DEFINITION 1.2. A *bilinear triple* (BT) is a set of three Banach spaces E, F, and G with a bilinear continuous mapping $B: E \times F \to G$ whose norm is ≤ 1; we write $x \cdot y = B(x, y)$ and denote the BT by (E, F, G).

Examples. Let W, X, and Y denote Banach spaces:

1. $E = L(X, Y)$, $F = L(W, X)$, $G = L(W, Y)$, and $B(v, u) = v \circ u$.

2. $E = L(X, Y)$, $F = X$, $G = Y$, and $B(u, x) = u(x)$; 2 is a particular case of 1: take $W = \mathbb{C}$.

3. $E = X'$, $F = X$, $G = \mathbb{C}$, and $B(x', x) = \langle x', x \rangle$; 3 is a particular case of 1: take $Y = W = \mathbb{C}$.

4. $E = G = Y$, $F = \mathbb{C}$, and $B(y, \lambda) = \lambda y$; 4 is a particular case of 1: take $X = W = \mathbb{C}$.

Given a BT (E, F, G) and functions $\alpha: [a, b] \to E$ and $f: [a, b] \to F$, for $(d, \xi) \in \mathcal{D}$ and $(d, \xi^\bullet) \in \mathcal{D}^\bullet$ we write respectively

$$\sigma_{d,\xi}(f; \alpha) = \sum_{i=1}^{|d|} \left[\alpha(t_i) - \alpha(t_{i-1})\right] \cdot f(\xi_i)$$

and

$$\sigma_{d,\xi^\bullet}(f;\alpha) = \sum_{i=1}^{|d|} \left[\alpha(t_i) - \alpha(t_{i-1})\right] \cdot f(\xi_i^\bullet)$$

and we define

$$\int_a^b d\alpha(t) \cdot f(t) = \lim_{\Delta d \to 0} \sigma_{d,\xi}(f;\alpha),$$

$$\oint_a^b d\alpha(t) \cdot f(t) = \lim_{d \in D} \sigma_{d,\xi}(f;\alpha),$$

$$\int_a^{\bullet b} d\alpha(t) \cdot f(t) = \lim_{d \in D} \sigma_{d,\xi^\bullet}(f;\alpha),$$

when these limits exist.

The first two integrals generalize the usual Riemann-Stieltjes integral (see [5] and also [6, 2, 4]) and the third one generalizes the Dushnik or *interior integral* (see [7, p. 96] and [3, 4]).

Given a function $\alpha: [a, b] \to E$ we denote by $R_\alpha([a, b], F)$, $R_\alpha^-([a, b], F)$, and $R_\alpha^\bullet([a, b], F)$ the vector spaces of all functions $f: [a, b] \to F$ such that there exist, respectively, the integrals $\int_a^b d\alpha(t) \cdot f(t)$, $\oint_a^b d\alpha(t) \cdot f(t)$, and $\int_a^{\bullet b} d\alpha(t) \cdot f(t)$. For $\alpha(t) \equiv t$ we write $R([a, b], F)$ etc.

By Proposition 1.1 we have $R_\alpha([a, b], F) \subset R_\alpha^-([a, b], F) \subset R_\alpha^\bullet([a, b], F)$ with the equality of the corresponding integrals.

Given $d \in D$, *the existence of the integrals*

$$\int_{t_{i-1}}^{\bullet t_i} d\alpha(t) \cdot f(t), \quad i = 1, 2, \ldots, |d|,$$

implies the existence of

$$\int_a^{\bullet b} d\alpha(t) \cdot f(t) = \sum_{i=1}^{|d|} \int_{t_{i-1}}^{\bullet t_i} d\alpha(t) \cdot f(t);$$

for the integral $\oint_a^b$ an analogous result is not true; on the other hand, *for the interior integral $\int_a^{\bullet b}$ the integration by parts formula is not valid,* but we have

THEOREM 1.2. Let (E, F, G) be a BT and $\alpha: [a, b] \to E$, $f: [a, b] \to F$ be bounded functions; if α or f is continuous, the existence of $\int_a^b d\alpha(t) \cdot f(t)$ implies the existence of the other two integrals and the three coincide (see [4, Theorem I.1.2]).

More generally we have:

THEOREM 1.2$\overline{}$. Let (E, F, G) be a BT and $\alpha: [a, b] \to E$, $f: [a, b] \to F$, bounded functions such that there exists $f_a^b\, d\alpha(t) \cdot f(t)$; if α and f have no common points of discontinuity then there exists

$$\int_a^b d\alpha(t) \cdot f(t) = f_a^b d\alpha(t) \cdot f(t).$$

THEOREM 1.2$^\bullet$. Let (E, F, G) be a BT and $\alpha: [a, b] \to E$, $f: [a, b] \to F$, bounded functions such that there exists $\int_{\bullet a}^b d\alpha(t) \cdot f(t)$; if α and f have no common points of left or right discontinuity then there exists

$$f_a^b d\alpha(t) \cdot f(t) = \int_{\bullet a}^b d\alpha(t) \cdot f(t).$$

In this chapter in general we give only results for the integrals $\int_a^b$ and $\int_{\bullet a}^b$; the corresponding theorems for f_a^b, and the proofs, are then obvious. Also, the proofs of the results for $\int_a^b$ in general follow from the corresponding results for $\int_{\bullet a}^b$, applying Theorem 1.2, and often they are much simpler than the direct proofs which do not use the interior integral.

Given $f \in R_\alpha([a, b], F)$, for every $(d, \xi) \in \mathcal{D}$ we define

$$f_{d,\xi} = f(\xi_1)\, \chi_{[a, t_1]} + \sum_{i=2}^{|d|} f(\xi_i)\, \chi_{]t_{i-1}, t_i]};$$

we have

$$\sigma_{d,\xi}(f;\, \alpha) = \int_{\bullet a}^b d\alpha(t) \cdot f_{d,\xi}(t)$$

and

$$\int_a^b d\alpha(t) \cdot f(t) = \lim_{\Delta d \to 0} \int_a^{b\bullet} d\alpha(t) \cdot f_{d,\xi}(t).$$

More generally we have

PROPOSITION 1.3. If $f \in R_\alpha([a, b], F)$, for every $t \in [a, b]$ we have

$$\int_a^t d\alpha(s) \cdot f(s) = \lim_{\Delta d \to 0} \int_a^{t\bullet} d\alpha(s) \cdot f_{d,\xi}(s).$$

Given $f \in R_\alpha^\bullet([a, b], F)$, for every $(d, \xi^\bullet) \in \mathcal{D}^\bullet$ we define

$$f_{d,\xi^\bullet} = f(\xi_1^\bullet) \chi_{[a,t_1]} + \sum_{i=2}^{|d|} f(\xi_i^\bullet) \chi_{]t_{i-1},t_i]};$$

we have

$$\sigma_{d,\xi^\bullet}(f; \alpha) = \int_a^{b\bullet} d\alpha(t) \cdot f_{d,\xi^\bullet}(t)$$

and

$$\int_a^{b\bullet} d\alpha(t) \cdot f(t) = \lim_{d \in D} \int_a^{b\bullet} d\alpha(t) \cdot f_{d,\xi^\bullet}(t).$$

In an analogous way we have

PROPOSITION 1.3$^\bullet$. If $f \in R_\alpha^\bullet([a, b], F)$, for every $t \in [a, b]$ we have

$$\int_a^{t\bullet} d\alpha(s) \cdot f(s) = \lim_{d \in D} \int_a^{t\bullet} d\alpha(s) \cdot f_{d,\xi^\bullet}(s).$$

DEFINITION 1.3. Given a parallelotope $I = \Pi_{1 \leq j \leq n} [a_j, b_j] \subset \mathbb{R}^n$ and a Banach space X, E(I, X) denotes the vector space of all finite linear combinations of characteristic functions of parallelotopes $\Pi_{1 \leq j \leq n} [c_j, d_j]$ contained in I (*step functions*). G(I, X) denotes the vector space of all *regulated functions*, i.e., functions that are uniform limits of step functions. Endowed with the norm $g \in G(I, X) \longmapsto \|g\| = \sup_{t \in I} \|g(t)\|$, G(I, X) is a Banach space.

It is not difficult to prove the following theorem.

THEOREM 1.4. For f: [a, b] → X the following properties are equivalent:

(a) $f \in G([a, b], X)$.

(b) For every $t \in [a, b[$ there exists $f(t+)$ and for every $t \in]a, b]$ there exists $f(t-)$.

(c) For every $\varepsilon > 0$ there exists $d \in D$ such that $\omega_d^{\cdot}(f) < \varepsilon$. See, for instance, [4], Theorem I.3.1.

We say that functions $f, g \in G([a, b], X)$ are equivalent if for every $t \in]a, b[$ we have $f(t+) = g(t+)$ [and hence also $f(t-) = g(t-)$]; we denote by $\tilde{G}([a, b], X)$ the Banach space of these equivalence classes.

PROPOSITION 1.5. Given a BT (E, F, G), $f \in G([a, b], E)$ and $g \in G([a, b], F)$, we have $f \cdot g \in G([a, b], G)$.

DEFINITION 1.4. Given a BT (E, F, G) and $\alpha: [a, b] \to E$ we define the B *variation* of α (where B denotes the bilinear mapping of the BT) by $SB[\alpha] = \sup_{d \in D} SB_d[\alpha]$, where

$$SB_d[\alpha] = \sup \left\{ \left\| \sum_{i=1}^{|d|} \left(\alpha(t_i) - \alpha(t_{i-1}) \right) \cdot y_i \right\| \mid y_i \in F, \|y_i\| \leq 1 \right\}.$$

DIRICHLET AND SUBSTITUTION FORMULAS

$SB([a, b], E)$ denotes the vector space of all functions of "bounded" (i.e., finite) B variation.

Examples.

1. In the case of Examples 1 and 2 of a BT we obtain the same space (see [4, Theorem I.4.4]), which we denote by $SV([a, b], L(X, Y))$, and we call its elements *functions of bounded semivariation*; in this case $SV[\alpha] = SB[\alpha]$ is called the *semivariation* of α.

2. In the case of Example 3 of a BT we obtain the space $BV([a, b], X')$ of functions of bounded variation and we have $V[\alpha] = SB[\alpha]$, where $V[\alpha] = V_{[a,b]}[\alpha]$ denotes the *variation* of α (on $[a, b]$): $V[\alpha] = \sup_{d \in D} V_d[\alpha]$ with $V_d[\alpha] = \Sigma_{i=1}^{|d|} \|\alpha(t_i) - \alpha(t_{i-1})\|$.

3. In the case of Example 4 of a BT we obtain the space $BW([a, b], Y)$ of *functions of weak bounded variation*; in this case $W[\alpha] = SB[\alpha]$ is called the *weak variation* of α. See [2, 4, 5, or 6].

THEOREM 1.6. Let (E, F, G) be a BT and $\alpha \in SB([a, b], E)$;

(a) For every $f \in C([a, b], F)$ there exists $F_\alpha[f] = \int_a^b d\alpha(t) \cdot f(t)$; we have $F_\alpha \in L[C([a, b], F), G]$ with $\|F_\alpha\| \leq SB[\alpha]$.

(b) For $(d, \xi) \in \mathcal{D}$ we have $\|\int_a^b d\alpha(t) \cdot f(t) - \sigma_{d,\xi}(f; \alpha)\| \leq SB[\alpha] \, \omega_{\Delta d}(f)$.

(See [2, Theorem I.1.5].)

THEOREM 1.6'. Let (E, F, G) be a BT and $\alpha \in SB([a, b], E)$;

(a) For every $f \in G([a, b], F)$ there exists $F_\alpha[f] = \int_a^b d\alpha(t) \cdot f(t)$; we have $\|F_\alpha[f]\| \leq SB[\alpha] \|f\|$.

(b) $F_\alpha[f]$ depends only on the class of f in $\tilde{G}([a, b], F)$ and hence $F_\alpha \in L[\tilde{G}([a, b], X), Y]$ with $\|F_\alpha\| \leq SB[\alpha]$.

(c) For every $(d, \xi^\cdot) \in \mathcal{D}^\cdot$ we have $\|\int_a^b d\alpha(t) \cdot f(t) - \sigma_{d,\xi^\cdot}(f; \alpha)\| \leq SB[\alpha] \, \omega_d^\cdot(f)$.

See [4, Theorem I.4.12].

DEFINITION 1.5. A Banach space system with an associative system of bilinear continuous mappings, or shortly, a *bilinear associative system* (BAS) is a system of six Banach spaces E_1, E_2, E_3, E_{12}, E_{23}, and E_{123} with four bilinear continuous mappings $B^{12}: E_1 \times E_2 \to E_{12}$, $B^{23}: E_2 \times E_3 \to E_{23}$, $B^{12,3}: E_{12} \times E_3 \to E_{123}$, $B^{1,23}: E_1 \times E_{23} \to E_{123}$ such that for every $x \in E_1$, $y \in E_2$, and $z \in E_3$ we have the "associative" relation $B^{1,23}[x, B^{23}(y, z)] = B^{12,3}[B^{12}(x, y), z]$. Unless otherwise stated we suppose that the bilinear mappings have norm ≤ 1; we write $x \cdot y = B^{12}(x, y)$, $x \cdot yz = B^{1,23}[x, B^{23}(y, z)]$, etc., and we write $(E_1, \ldots, E_{123})$ to denote the BAS.

Examples.

1. W, X, Y, and Z are Banach spaces; $E_1 = L(Y, Z)$, $E_2 = L(X, Y)$, $E_3 = L(W, X)$, $E_{12} = L(X, Z)$, $E_{23} = L(W, X)$, and $E_{123} = L(W, Z)$; the bilinear mappings are the natural compositions.

2. X, Y, and Z are Banach spaces; $E_1 = L(Y, Z)$, $E_2 = L(X, Y)$, $E_3 = X$, $E_{12} = L(X, Z)$, $E_{23} = Y$, and $E_{123} = Z$; the bilinear mappings are the natural ones; 2 is a particular case of 1: take $W = \mathbb{C}$ etc.

3. All spaces are equal to a Banach algebra A [for instance, $L(X)$] the bilinear mapping being the product.

4. A BT (E, F, G) is a particular case of a BAS: we take $E_1 = E$, $E_2 = \mathbb{C}$, $E_3 = F$, $E_{12} = E$, $E_{23} = F$, and $E_{123} = G$; $B^{12}(x, \lambda) = \lambda x$, $B^{23}(\lambda, y) = \lambda y$, $B^{1,23} = B = B^{12,3}$.

DIRICHLET AND SUBSTITUTION FORMULAS

5. If (E, F, G) is a BT we may define the following BAS: $E_1 = E$, $E_2 = F$, $E_3 = G'$, $E_{12} = G$, $E_{23} = E'$, $E_{123} = \mathbb{C}$ and $B^{12}(x, y) = x \cdot y$; $B^{23}(y, z')$ is the element $x' \in E'$ such that $\langle x, x' \rangle = \langle xy, z' \rangle$ for every $x \in E$ (since the linear form $x \in E \mapsto \langle xy, z' \rangle \in \mathbb{C}$ is continuous, x' is well-defined); we write $x' = yz' = B^{23}(y, z')$; $B^{12,3}(z, z') = \langle z, z' \rangle$, $B^{1,23}(x, x') = \langle x, x' \rangle$; it is immediate that $\|yz'\| \leq \|y\|\|z'\|$, i.e., $\|B^{23}\| \leq 1$.

Remark. For certain theorems one needs systems with four or even five base spaces E_i. The definitions in these cases are obvious; in the case $n = 4$, for instance, we write the BAS $(E_1, \ldots, E_{1234})$; it is formed by ten spaces, eight bilinear mappings, and four "associative" equalities.

THEOREM 1.7. Let $(E_1, \ldots, E_{123})$ be a BAS that satisfies the following property:

(*) For every $u \in E_{12}$ we have $\|u\| = \sup\{\|B^{12,3}(u, z)\| \mid \|z\| \leq 1\}$.

Then if $\alpha \in SB^{1,23}([a, b], E_1)$, we have $\alpha \in SB^{12}([a, b], E_2)$ and $SB^{12}[\alpha] \leq SB^{1,23}[\alpha]$.

Proof. For every $d \in D$ we have

$$SB_d^{12}[\alpha]$$
$$= \sup\left\{ \left\| \sum_{i=1}^{|d|} \left(\alpha(t_i) - \alpha(t_{i-1})\right) \cdot x_i^2 \right\| \mid x_i^2 \in E_2, \|x_i^2\| \leq 1 \right\}$$
$$= \sup\left\{ \sup_{\|x^3\| \leq 1} \left\| \left[\sum_{i=1}^{|d|} \left(\alpha(t_i) - \alpha(t_{i-1})\right) x_i^2 \right] x^3 \right\| \mid x_i^2 \in E_2, \|x_i^2\| \leq 1 \right\}$$
$$\leq \sup\left\{ \left\| \sum_{i=1}^{|d|} \left(\alpha(t_i) - \alpha(t_{i-1})\right) x_i^{23} \right\| \mid x_i^{23} \in E_{23}, \|x_i^{23}\| \leq 1 \right\}$$
$$= SB_d^{1,23}[\alpha].$$

The notion of BAS allows us to unify in a natural way many different situations. Let us also mention that all results of Chapter II of [4], particularly the Dirichlet and substitution formulas, extended trivially to BAS that satisfy the property (*) of Theorem 1.7.

There are also results for BT whose proof is very much simplified if we consider them as part of an adequate BAS. For instance, all results of Chapter II of [4] are still true if instead of functions of bounded semivariation (and the semi-variation norm) we consider functions of bounded variation (and the variation norm). The verification of this assertion is trivial but for Lemma 1.2, whose proof does not change if we use an adequate BAS (see Theorem 1.8 that follows).

THEOREM 1.8. Let (E, F, G) be a BT, $\alpha \in BV([c, d], E)$, $h: [c, d] \times [a, b] \to F$ a function that is regulated as a function of the first variable and that is uniformly of bounded variation as a function of the second variable (i. e., for every $t \in [c,d]$ we have $h^t \in BV([a, b], F)$ and $\sup_{c \leq t \leq d} V[h^t] < \infty$); for every $s \in [a, b]$ we define $\bar{h}(s) = \int_c^d d\alpha(t) \cdot h(t, s)$. Then we have $\bar{h} \in BV([a, b], G)$ and $V[\bar{h}] \leq V[\alpha] \sup_{c \leq t \leq d} V[h^t]$.

Proof. By Theorem 1.6', $\bar{h}$ is well-defined (since $SB[\alpha] \leq V[\alpha]$). Let us take the BAS associated to the BT (E, F, G) as in Example 5. It is immediate that for $d \in D_{[a,b]}$ we have

$V_d[\bar{h}]$

$= \sum_{i=1}^{|d|} \|\bar{h}(s_i) - \bar{h}(s_{i-1})\| = \sup_{\|z_i'\| \leq 1} \sum_{i=1}^{|d|} |<\bar{h}(s_i) - \bar{h}(s_{i-1}), z_i'>|$

$= \sup_{\|z_i'\| \leq 1} \left| \sum_{i=1}^{|d|} <\bar{h}(s_i) - \bar{h}(s_{i-1}), z_i'> \right|$ where $z_i' \in G'$.

DIRICHLET AND SUBSTITUTION FORMULAS

Given $z_i' \in G'$ with $\|z_i'\| \leq 1$, $i = 1, 2, \ldots, |d|$, we have

$$\left| \sum_{i=1}^{|d|} <\bar{h}(s_i) - \bar{h}(s_{i-1}), z_i'> \right| =$$

$$= \left| \sum_{i=1}^{|d|} <\int_c^d \cdot d\alpha(t) \cdot \left(h(t, s_i) - h(t, s_{i-1})\right), z_i'> \right|$$

$$= \left| \sum_{i=1}^{|d|} \int_c^d <d\alpha(t), \left(h(t, s_i) - h(t, s_{i-1})\right) z_i'> \right|$$

$$= \left| \int_c^d <d\alpha(t), \sum_{i=1}^{|d|} \left(h(t, s_i) - h(t, s_{i-1})\right) z_i'> \right|$$

$$\leq V[\alpha] \sup_{c \leq t \leq d} \left\| \sum_{i=1}^{|d|} \left(h(t, s_i) - h(t, s_{i-1})\right) z_i' \right\|$$

$$\leq V[\alpha] \sup_{c \leq t \leq d} \sum_{i=1}^{|d|} \|h(t, s_i) - h(t, s_{i-1})\| \leq V[\alpha] \sup_{c \leq t \leq d} V[h^t],$$

hence the result.

<u>Remark</u>. By the application of (c) of Theorem 1.6˙ one may also prove Theorem 1.8 without the consideration of BAS.

2. THE DIRICHLET FORMULA

<u>THEOREM 2.1˙</u>. Let $(E_1, \ldots, E_{123})$ be a BAS that satisfies the property (*) of Theorem 1.7; for $\alpha \in SB^{1,23}([a, b], E_1)$, $h \in G([a, b] \times [a, b], E_2)$, and $\beta \in BV\,C([a, b], E_3) = BV([a, b], E_3) \cap C([a, b], E_3)$ we have

$$(D\dot{}) \qquad \int_a^b \left[\int_a^s \cdot d\alpha(t) \cdot h(t, s)\right] d\beta(s)$$

$$= \int_a^b \cdot d\alpha(t) \left[\int_t^b h(t, s)\, d\beta(s)\right].$$

Proof. We will show that both members of (D·) are well defined and that they depend continuously on h; thereafter it will be enough to prove the equality when h is a step function.

1a. We define $\bar{h}(t) = \int_t^b h(t, s) \, d\beta(s)$; by Theorem 1.6· and by Theorem 1.2, for every $t \in [a, b]$ this integral is well defined and we have $\|\bar{h}(t)\| \le \|h\| \, V[\beta]$. Therefore the mapping

$$h \in G([a, b] \times [a, b], E_2) \longmapsto \bar{h} \in B([a, b], E_{23})$$

is continuous. ($B([a, b], X)$ denotes the Banach space of all *bounded* functions $f: [a, b] \to X$). Let us prove that $\bar{h} \in G([a, b], E_{23})$; it is enough to do it when h is a step function because the general case follows by uniform convergence. Any step function is a finite linear combination of functions of the form $h = \chi_{|a,c|} \otimes \chi_{|a,d|} y$, and for this function we have

$$\bar{h}(t) = \chi_{|a,c|}(t) \int_t^b \chi_{|a,d|}(s) \, y \, d\beta(s)$$

$$= \chi_{|a,c|}(t) \, y \Big(\beta(b) - \beta(t \vee d) \Big),$$

which is really regulated.

1b. The mapping $g \in G([a, b], E_{23}) \longmapsto \int_a^b d\alpha(t) \cdot g(t) \in E_{123}$ is well defined and continuous, since $\alpha \in SB^{1,23}([a, b], E_1)$; hence the composed mapping

$$h \in G([a, b] \times [a, b], E_2) \longmapsto \int_a^b d\alpha(t) \left[\int_t^b h(t, s) \, d\beta(s) \right] \in E_{123}$$

is well defined and continuous.

Remark. 1a. and 1b remain true if we suppose only $\beta \in BV([a, b], E_3)$ and replace $\int_t^b$ by $\int_t^b \cdot$.

2a. We define $\tilde{h}(s) = \int_a^s d\alpha(t) \cdot h(t, s)$; by Theorem 1.6· this integral is well defined, since $\alpha \in SB^{12}([a, b], E_1)$ by

DIRICHLET AND SUBSTITUTION FORMULAS

Theorem 1.7. We have $\|\tilde{h}(s)\| \leq \|h\|\ SB^{12}[\alpha] \leq \|h\|\ SB^{1,23}[\alpha]$ and therefore the linear mapping

$$h \in G([a, b] \times [a, b], E_2) \longmapsto \tilde{h} \in B([a, b], E_{12})$$

is continuous.

2b. We will now prove that for every $h \in G([a, b] \times [a, b], E_2)$ the integral $\int_a^b \tilde{h}(s)\ d\beta(s)$ is well defined and depends continuously on h. Let $h_n \in E([a, b] \times [a, b], E_2)$ be a sequence of step functions that converges uniformly to h. Let us first prove that there exists

$$\int_a^b \tilde{h}_n(s)\ d\beta(s) = \int_a^b \left[\int_a^s d\alpha(t) \cdot h_n(t, s)\right] d\beta(s).$$

It is enough to show it for $h_n = \chi_{|a,c|} \otimes \chi_{|a,d|} y$; in this case we have

$$\int_a^b \chi_{|a,d|}(s) \left[\int_a^s d\alpha(t) \cdot y\ \chi_{|a,c|}(t)\right] d\beta(s)$$

$$= \int_a^b \chi_{|a,d|}(s) \left[\alpha(s \wedge c) - \alpha(a)\right] \cdot y\ d\beta(s)$$

$$= \int_a^d \left[\alpha(s \wedge c) - \alpha(s)\right] \cdot y\ d\beta(s),$$

and β *being continuous* we see, *using integration by parts*, that this integral exists because $\int_a^{c \wedge d} d\alpha(s) \cdot y\beta(s)$ exists.

From 2a it follows that $\tilde{h}_n$ tends uniformly to h [and that the sequence $\int_a^b \tilde{h}_n(s)\ d\beta(s)$ converges]. We will now prove that the integral $\int_a^b \tilde{h}(s)\ d\beta(s)$ exists: we have to show that for every $\varepsilon > 0$ there exists a $\delta > 0$ such that for $d, \overline{d} \in D$ with $\Delta d, \Delta \overline{d} < \delta$ we have

$$\|\sigma_{d,\xi}(\tilde{h};\ \beta) - \sigma_{\overline{d},\overline{\xi}}(\tilde{h};\ \beta)\| < \varepsilon.$$

We have

$$\|\sigma_{d,\xi}(\tilde{h}; \beta) - \sigma_{\overline{d},\overline{\xi}}(\tilde{h}; \beta)\|$$

$$\leq \|\sigma_{d,\xi}(\tilde{h} - \tilde{h}_n; \beta) - \sigma_{\overline{d},\overline{\xi}}(\tilde{h} - \tilde{h}_n; \beta)\|$$

$$+ \|\sigma_{d,\xi}(\tilde{h}_n; \beta) - \sigma_{\overline{d},\overline{\xi}}(\tilde{h}_n; \beta)\|.$$

The first summand is $\leq 2\|\tilde{h} - \tilde{h}_n\| V[\beta]$ and becomes $< \varepsilon/2$ for n sufficiently large because $\tilde{h}_n$ converges uniformly to $\tilde{h}$; if we fix such an n, there exists a $\delta > 0$ such that for $\Delta d, \Delta \overline{d} < \delta$ the second summand becomes smaller than $\varepsilon/2$ because we already proved that there exists $\int_a^b \tilde{h}_n(s) \, d\beta(s)$.

From $\|\int_a^b \tilde{h}(s) \, d\beta(s)\| \leq \|\tilde{h}\| V[\beta]$ it follows that the integral depends continuously on $\tilde{h}$; by 2a it follows, through compositions, that the linear mapping

$$h \in G([a, b] \times [a, b], E_2) \longmapsto \int_a^b \left[\int_a^s \bullet \, d\alpha(t) \bullet h(t, s) \right] d\beta(s) \in E_{123}$$

is continuous.

3. By our initial remarks it is now enough to prove the Dirichlet formula (D·) for $h = \chi_{|a,c|} \otimes \chi_{|a,d|} y$. We have

$$\int_a^b \left[\int_a^s \bullet \, d\alpha(t) \bullet y \, \chi_{|a,c|}(t) \, \chi_{|a,d|}(s) \right] d\beta(s)$$

$$= \int_a^{c \wedge d} \left[\int_a^s \bullet \, d\alpha(t) \bullet y \right] d\beta(s)$$

$$= \int_a^{c \wedge d} \left(\alpha(s) - \alpha(a) \right) d\beta(s)$$

and

$$\int_a^b \bullet \, d\alpha(t) \left[\int_t^b \chi_{|a,c|}(t) \, \chi_{|a,d|}(s) \, y \, d\beta(s) \right]$$

$$= \int_a^c d\alpha(t) \bullet y \left(\beta(c \wedge d) - \beta(t \wedge d) \right)$$

$$= \int_a^{c \wedge d} d\alpha(t) \bullet y \left(\beta(c \wedge d) - \beta(t) \right),$$

and the equality of these two integrals follows through integration by parts.

DIRICHLET AND SUBSTITUTION FORMULAS

Remark. Theorem 2.1˙ remains true for $\beta \in BV([a, b], E_3)$ if we suppose that α is also regulated and if we replace $\int$ by $\int \cdot$ (under these hypothesis $\tilde{h}$ is regulated: it is enough to prove it for $h = \chi_{]a,c|} \otimes \chi_{|a,d|} y$, and this is immediate).

Without these restrictions the Dirichlet formula may not be valid.

Example. We take $\alpha \in BW([a, b], L_\infty([a, b]))$ defined by $\alpha(t) = \chi_{[a,t]}$ and $\beta = \chi_{[c,d]} \in BV([a, b])$, where $a < c \leq b$. We do not have $\int_a^b \cdot [\int_a^s d\alpha(t)] \, d\beta(s) = \int_a^b \cdot d\alpha(t) \int_t^b d\beta(s)$, i.e., $\int_a^b \cdot [\alpha(s) - \alpha(a)] \, d\beta(s) = \int_a^b d\alpha(t) [\beta(b) - \beta(t)]$ because $\int_a^b \cdot \alpha(s) \, d\beta(s)$ does not exist: it should be $\alpha(c-)$, and this limit does not exist since for $t' < t'' < c$ we have $\|\alpha(t'') - \alpha(t')\| = \|\chi_{]t',t'']}\| = 1$.

By the way, if the integration by parts formula

$$\int_a^b \cdot d\alpha(t) \cdot \beta(t) = \alpha(b) \cdot \beta(b) - \alpha(a) \cdot \beta(a) - \int_a^b \cdot \alpha(t) \cdot d\beta(t)$$

is not valid for two functions α and β, then neither is the Dirichlet formula

$$\int_a^b \cdot \left[\int_a^s d\alpha(t) \right] d\beta(s) = \int_a^b \cdot d\alpha(t) \int_t^b d\beta(s)$$

that is equivalent to it.

THEOREM 2.1. Let $(E_1, \ldots, E_{123})$ be a BAS that satisfies the property (*) of Theorem 1.7; for $\alpha \in SB^{1,23}([a, b], E_1)$, $h \in C([a, b] \times [a, b], E_2)$, and $\beta \in BV \, C([a, b], E_3)$ we have

(D) $$\int_a^b \left[\int_a^s d\alpha(t) \cdot h(t, s) \right] d\beta(s)$$

$$= \int_a^b d\alpha(t) \left[\int_t^b h(t, s) \, d\beta(s) \right].$$

Proof. The proof follows immediately from Theorem 2.1˙ by Theorem 1.2 (since under the hypothesis of Theorem 2.1 the function $t \in [a, b] \longmapsto \int_t^b h(t, s)\, d\beta(s) \in E_{23}$ is continuous, as it is easy to see).

Remark. (D) remains true for $\beta \in BV([a, b], E_3)$ if α is also continuous, as follows from the remark after Theorem 2.1˙ (under these hypothesis $\tilde{h}$ is continuous).

THEOREM 2.2˙. Let $(E_1, \ldots, E_{123})$ be a BAS that satisfies the property (*) of Theorem 1.7; for $\alpha \in SB^{1,23}([a, b], E_1)$, $h \in G([a, b], E_2)$ and $\beta \in BV\, C([a, b], E_3)$ we have

$$\int_a^b d_s \left[\int_a^s {}^\bullet d\alpha(t) \cdot h(t) \right] \beta(s) = \int_a^b {}^\bullet d\alpha(s) \cdot h(s)\, \beta(s).$$

Proof. By Theorem 2.1˙ we have

$$\int_a^b \left[\int_a^s {}^\bullet d\alpha(t) \cdot h(t) \right] d\beta(s) = \int_a^b {}^\bullet d\alpha(t) \left[\int_t^b h(t)\, d\beta(s) \right].$$

Integrating by parts we get

$$\int_a^b {}^\bullet d\alpha(t) \cdot h(t)\, \beta(b) - \int_a^b d_s \left[\int_a^s {}^\bullet d\alpha(t) \cdot h(t) \right] \beta(s)$$

$$= \int_a^b {}^\bullet d\alpha(t) \cdot h(t) \left(\beta(b) - \beta(t) \right)$$

and hence the result.

THEOREM 2.2. Let $(E_1, \ldots, E_{123})$ be a BAS that satisfies the property (*) of Theorem 1.7; for $\alpha \in SB^{1,23}([a, b], E_1)$, $h \in C([a, b], E_2)$, and $\beta \in BV\, C([a, b], E_3)$ we have

$$\int_a^b d_s \left[\int_a^s d\alpha(t) \cdot h(t) \right] \beta(s) = \int_a^b d\alpha(s) \cdot h(s)\, \beta(s).$$

Proof. The proof follows from Theorem 2.2˙ by Theorem 1.2.

Along the same line we have the following results for iterated integrals.

THEOREM 2.3˙. Let $(E_1, \ldots, E_{123})$ be a BAS that satisfies the property (*) of Theorem 1.7; for $\alpha \in SB^{1,23}([c, d], E_1)$, $h \in G([c, d] \times [a, b], E_2)$, and $\beta \in BV([a, b], E_3)$ we have

$$\int_a^b \left[\int_c^d d\alpha(t) \cdot h(t, s) \right] d\beta(s) = \int_c^d d\alpha(t) \left[\int_a^b h(t, s) \, d\beta(s) \right].$$

The proof follows the steps of the proof of Theorem 2.1˙ with some simplifications, mainly in the part corresponding to 2b since now the function $s \in [a, b] \longmapsto \int_c^d d\alpha(t) \cdot h(t, s) \in E_{12}$ is regulated.

Remark. Theorem 2.1˙ cannot be deduced from Theorem 2.3˙ since the function $(t, s) \in [a, b] \times [a, b] \longmapsto Y(s - t)h(t, s) \in E_2$ is, in general, not regulated!

THEOREM 2.3. Let $(E_1, \ldots, E_{123})$ be a BAS that satisfies the property (*) of Theorem 1.7; for $\alpha \in SB^{1,23}([c, d], E_1)$, $h \in C([c, d] \times [a, b], E_2)$ and $\beta \in BV([a, b], E_3)$ and we have

$$\int_a^b \left[\int_c^d d\alpha(t) \cdot h(t, s) \right] d\beta(s) = \int_c^d d\alpha(t) \left[\int_a^b h(t, s) \, d\beta(s) \right].$$

Proof. Since the functions

$$s \in [a, b] \longmapsto \int_c^d d\alpha(t) \cdot h(t, s) \in E_{12}$$

and

$$t \in [c, d] \longmapsto \int_a^b h(t, s) \, d\beta(s) \in E_{23}$$

are continuous the result follows from Theorem 2.3˙ by Theorem 1.2.

3. THE SUBSTITUTION FORMULA (I)

The purpose of this section is to prove the following versions of the substitution formula:

THEOREM 3.6˙. Let $(E_1, \ldots, E_{123})$ be a BAS, $\alpha \in SB^{1,23}([a,b], E_1)$, $f \in R_\alpha^{\cdot}([a, b], E_2)$, and $g \in G([a, b], E_3)$; we have

$$(S^{\cdot}) \qquad \int_a^b {}_\cdot d_t \left[\int_a^t {}_\cdot d\alpha(s) \cdot f(s) \right] g(t) = \int_a^b {}_\cdot d\alpha(t) \cdot f(t) \, g(t).$$

THEOREM 3.6. Let $(E_1, \ldots, E_{123})$ be a BAS, $\alpha \in SB^{1,23}([a,b], E_1)$, $f \in R_\alpha([a, b], E_2)$, and $g \in C([a, b], E_3)$; we have

$$(S) \qquad \int_a^b d_t \left[\int_a^t d\alpha(s) \cdot f(s) \right] g(t) = \int_a^b d\alpha(t) \cdot f(t) \, g(t).$$

In order to prove these theorems we need several preliminary results.

THEOREM 3.1˙. Under the hypothesis of Theorem 3.6˙ we have

(a) $I_f \in SB^{12,3}([a, b], E_{12})$ where $I_f(t) = \int_a^t {}_\cdot d\alpha(s) \cdot f(s)$.

(b) $SB^{12,3}[I_f] \leq \|f\| \, SB^{1,23}[\alpha]$.

(c) There exists $\int_a^b {}_\cdot d_t \left[\int_a^t {}_\cdot d\alpha(s) \cdot f(s) \right] g(t)$.

DIRICHLET AND SUBSTITUTION FORMULAS

Proof. We first prove (b). Given $d \in D$ and $z_i \in E_3$ with $\|z_i\| \leq 1$, we have

$$\left\| \sum_{i=1}^{|d|} [I_f(t_i) - I_f(t_{i-1})] z_i \right\|$$

$$= \left\| \sum_{i=1}^{|d|} \int_{t_{i-1}}^{t_i} \cdot \, d\alpha(t) \cdot f(t) \, z_i \right\|$$

$$= \left\| \int_a^b \cdot \, d\alpha(t) \cdot f_z(t) \right\|$$

$$\leq \|f_z\| \, SB^{1,23}[\alpha],$$

since we have obviously $f_z \in R_\alpha^\cdot([a, b], E_3)$, where $f_z(t) = f(t) z_i$ for $t \in \,]t_{i-1}, t_i[$; hence the result, since $\|f_z\| \leq \|f\|$. Part (a) follows from (b), and (c) follows from (a).

By Theorem 1.2, Theorem 3.1˙ implies the following:

THEOREM 3.1. Under the hypothesis of Theorem 3.6 we have

(a) $I_f \in SB^{12,3}([a, b], E_{12})$, where $I_f(t) = \int_a^t d\alpha(s) \cdot f(s)$.

(b) $SB^{12,3}[I_f] \leq \|f\| \, SB^{1,23}[\alpha]$.

(c) There exists $\int_a^b d_t [\int_a^t d\alpha(s) \cdot f(s)] \, g(t)$.

THEOREM 3.2˙. With the notations of Theorem 3.6˙, if $f \in G([a, b], E_2)$ and if $g: [a, b] \to E_3$ is such that one of the integrals in (S˙) exists, then so does the other and we have (S˙).

Proof. Let us take approximating sums for both integrals in (S˙); for $(d, \xi\dot{}) \in \mathcal{D}\dot{}$ we have

$$\left\| \sum_{i=1}^{|d|} \int_{t_{i-1}}^{t_i} d\alpha(s) f(s) \cdot g(\xi_i^\cdot) - \sum_{i=1}^{|d|} \left[\alpha(t_i) - \alpha(t_{i-1})\right] \cdot f(\xi_i^\cdot) g(\xi_i^\cdot) \right\|$$

$$= \left\| \sum_{i=1}^{|d|} \int_{t_{i-1}}^{t_i} d\alpha(s) \cdot \left[f(s) - f(\xi_i^\cdot)\right] g(\xi_i^\cdot) \right\|$$

$$\leq SB^{1,23}[\alpha] \, \omega_d^\cdot(f) \, \|g\|,$$

hence the result by (c) of Theorem 1.4.

THEOREM 3.2. With the notations of Theorem 3.6, if $f \in C([a, b], E_2)$ and if $g: [a, b] \to E_3$ is such that one of the integrals in (S) exists, then so does the other and we have (S).

Proof. Let us take approximating sums for both integrals in (S); for $(d, \xi) \in \mathcal{D}$ we have

$$\left\| \sum_{i=1}^{|d|} \int_{t_{i-1}}^{t_i} d\alpha(s) f(s) \cdot g(\xi_i) - \sum_{i=1}^{|d|} \left[\alpha(t_i) - \alpha(t_{i-1})\right] \cdot f(\xi_i) g(\xi_i) \right\|$$

$$= \left\| \sum_{i=1}^{|d|} \int_{t_{i-1}}^{t_i} d\alpha(s) \left[f(s) - f(\xi_i)\right] \cdot g(\xi_i) \right\|$$

$$\leq SB^{1,23}[\alpha] \, \omega_{\Delta d}(f) \, \|g\|;$$

hence the result because $\omega_{\Delta d}(f) \to 0$ if $\Delta d \to 0$, since f is uniformly continuous in $[a, b]$.

LEMMA 3.3·. Under the hypothesis of Theorem 3.6· (S·) is true if $f \in G([a, b], E_2)$.

Proof. By Proposition 1.5 the second integral in (S·) exists; by Theorem 3.1· the first one exists, hence the result by Theorem 3.2·.

DIRICHLET AND SUBSTITUTION FORMULAS

LEMMA 3.3. Under the hypothesis of Theorem 3.6 (S) is true if $f \in C([a, b], E_2)$.

Proof. The proof follows from Lemma 3.3˙ by Theorem 1.2.

PROPOSITION 3.4˙. Under the hypothesis of Theorem 3.6˙ we have

$$\int_a^b \cdot d_t \left[\int_a^t \cdot d\alpha(s) \cdot f(s) \right] g(t)$$

$$= \lim_{d \in D} \int_a^b \cdot d_t \left[\int_a^t \cdot d\alpha(s) \cdot f_{d,\xi\cdot}(s) \right] g(t).$$

Proof. $f_{d,\xi\cdot}$ has been defined in Sec. 1. By Theorem 3.1˙ the first integral exists; we define $I(t) = \int_a^t \cdot d\alpha(s) \cdot f(s)$ and $I_{(d,\xi\cdot)}(t) = \int_a^t \cdot d\alpha(s) \cdot f_{d,\xi\cdot}(s)$. For $(\bar{d}, \bar{\xi}\cdot) \in \mathcal{D}\cdot$ we have

$$\left\| \int_a^b \cdot d_t \left[\int_a^t \cdot d\alpha(s) \cdot f(s) \right] g(t) - \int_a^b \cdot d_t \left[\int_a^t \cdot d\alpha(s) \cdot f_{d,\xi\cdot}(s) \right] g(t) \right\|$$

$$= \left\| \int_a^b \cdot dI(t) \cdot g(t) - \int_a^b \cdot dI_{(d,\xi\cdot)}(t) \cdot g(t) \right\|$$

$$\leq \left\| \int_a^b \cdot dI_{(d,\xi\cdot)}(t) \cdot g(t) - \int_a^b \cdot d\alpha(t) \cdot f_{d,\xi\cdot}(t) g(t) \right\|$$

$$+ \left\| \int_a^b \cdot d\alpha(t) \cdot f_{d,\xi\cdot}(t) g(t) - \sum_{j=1}^{|\bar{d}|} \int_{\bar{t}_{j-1}}^{\bar{t}_j} \cdot d\alpha(t) \cdot f_{d,\xi\cdot}(t) g(\bar{\xi}\cdot_j) \right\|$$

$$+ \left\| \sum_{j=1}^{|\bar{d}|} \int_{\bar{t}_{j-1}}^{\bar{t}_j} \cdot d\alpha(t) \cdot f_{d,\xi\cdot}(t) g(\bar{\xi}\cdot_j) - \sum_{j=1}^{|\bar{d}|} \left\{ I_{(d,\xi\cdot)}(\bar{t}_j) - I_{(d,\xi\cdot)}(\bar{t}_{j-1}) \right\} g(\bar{\xi}\cdot_j) \right\|$$

$$+ \left\| \sum_{j=1}^{|\bar{d}|} \left\{ (I_{(d,\xi\cdot)}(\bar{t}_j) - I(\bar{t}_j)) - (I_{(d,\xi\cdot)}(\bar{t}_{j-1}) - I(\bar{t}_{j-1})) \right\} g(\bar{\xi}\cdot_j) \right\|$$

$$+ \left\| \sum_{j=1}^{|\bar{d}|} \left\{ I(\bar{t}_j) - I(\bar{t}_{j-1}) \right\} g(\bar{\xi}\cdot_j) - \int_a^b \cdot dI(t) \cdot g(t) \right\|.$$

By the definition of $I_{(d,\xi^*)}$ the third summand is zero; by Lemma 3.3* the first summand is zero, since $f_{d,\xi^*} \in G([a,b], E_2)$. By Theorem 3.1* (c) and by Theorem 1.6* (c) the fifth summand is $\leq \|f\| \, SB^{1,23}[\alpha] \, \omega_{\overline{d}}^{\cdot}(g)$ and the same is true for the second summand, *for all* $(d, \xi^*) \in \mathcal{D}^*$. Hence, given $\varepsilon > 0$, there exists $d_\varepsilon \in D$ such that for $\overline{d} \geq d_\varepsilon$ the second and the fifth summands are $\leq \varepsilon/3$ for all $d \in D$; we fix such a $\overline{d} \geq d_\varepsilon$; then, by Proposition 1.3, the fourth summand becomes $\leq \varepsilon/3$ for all d sufficiently "large."

PROPOSITION 3.4. Under the hypothesis of Theorem 3.6 we have

$$\int_a^b d_t \left[\int_a^t d\alpha(s) \cdot f(s) \right] g(t)$$

$$= \lim_{\Delta d \to 0} \int_a^b d_t \left[\int_a^t \cdot d\alpha(s) \cdot f_{d,\xi}(s) \right] g(t).$$

Proof. The proof follows the steps of the proof of the Proposition 3.4* if we take $\mathcal{D}$ instead of $\mathcal{D}^*$, $\lim_{\Delta d \to 0}$ instead of $\lim_{d \in D}$, $\omega_{\Delta \overline{d}}(g)$ instead of $\omega_{\overline{d}}^{\cdot}(g)$, Theorem 1.6 (b) instead of Theorem 1.6* (c), etc.

LEMMA 3.5*. Under the hypothesis of Theorem 2.6*, for all $(d, \xi^*) \in \mathcal{D}^*$ there exists $\int_a^b d\alpha(t) \cdot f_{d,\xi^*}(t) \, g(t)$ and we have

$$\left\| \int_a^b \cdot d\alpha(t) \cdot f_{d,\xi^*}(t) \, g(t) - \sum_{i=1}^{|d|} \left(\alpha(t_i) - \alpha(t_{i-1}) \right) \cdot f(\xi_i^*) \, g(\xi_i^*) \right\|$$

$$\leq SB^{1,23}[\alpha] \, \|f\| \, \omega_d^{\cdot}(g).$$

Proof. Since $f_{d,\xi^*} \in E([a,b], E_2) \subset G([a,b], E_2)$, the existence of the integral follows from Lemma 3.3*. The inequality is obvious.

In the same way one proves the following:

DIRICHLET AND SUBSTITUTION FORMULAS 159

LEMMA 3.5. Under the hypothesis of Theorem 3.6, for all $(d, \xi) \in \mathcal{D}$ there exists $\int_a^b d\alpha(t) \cdot f_{d,\xi}(t)\, g(t)$ and we have

$$\left\| \int_a^b d\alpha(t) \cdot f_{d,\xi}(t)\, g(t) - \sum_{i=1}^{|d|} \left(\alpha(t_i) - \alpha(t_{i-1}) \right) \cdot f(\xi_i)\, g(\xi_i) \right\|$$

$$\leq SB^{1,23}[\alpha]\, \|f\|\, \omega_{\Delta d}(g).$$

Proof of Theorem 3.6*. The first integral in (S*) exists by Theorem 3.1* and is, by Proposition 3.4*, equal to

$$\lim_{d \in D} \int_a^b d_t \left[\int_a^t d\alpha(s) \cdot f_{d,\xi}\cdot(s) \right] g(t),$$

which by Lemma 3.3* is equal to

$$\lim_{d \in D} \int_a^b d\alpha(t) \cdot f_{d,\xi}\cdot(t)\, g(t);$$

by Lemma 3.5* this limit is equal to

$$\lim_{d \in D} \sum_{i=1}^{|d|} \left(\alpha(t_i) - \alpha(t_{i-1}) \right) \cdot f(\xi_i^*)\, g(\xi_i^*)$$

and by definition this is

$$\int_a^b d\alpha(t) \cdot f(t)\, g(t).$$

Proof of Theorem 3.6. The proof follows the steps of the proof of Theorem 3.6* (applying Theorem 3.1, Proposition 3.4, etc. instead of Theorem 3.1*, Proposition 3.4*, etc.).

4. THE DARBOUX-STIELTJES INTEGRAL

If X is a Banach space and $\alpha \in BV([a, b], X)$, we define $\hat{\alpha} \in BV([a, b], \mathbb{R})$ by $\hat{\alpha}(t) = V_{[a,t]}[\alpha]$, $t \in [a, b]$; given $d \in D$, we write $V_i[\alpha] = V_{[t_{i-1},t_i]}[\alpha] = V_{[t_{i-1},t_i]}[\hat{\alpha}] = \hat{\alpha}(t_i) - \hat{\alpha}(t_{i-1})$, $i = 1, 2, \ldots, |d|$.

Let F be a Banach space and $f: [a, b] \to F$ a bounded function. We write $f \in D_\alpha^\cdot([a, b], F)$, and we say that f *satisfies the interior Darboux condition* (with respect to α) if

$$\lim_{d \in D} \sum_{i=1}^{|d|} V_i[\alpha] \, \omega_i^\cdot(f) = 0.$$

We have $D_\alpha^\cdot([a, b], F) = D_{\hat{\alpha}}^\cdot([a, b], F)$. We write $f \in D_\alpha([a, b], F)$, and we say that f *satisfies the Darboux condition* (with respect to α) if

$$\lim_{\Delta d \to 0} \sum_{i=1}^{|d|} V_i[\alpha] \, \omega_i(f) = 0.$$

We have $D_\alpha([a, b], F) = D_{\hat{\alpha}}([a, b], F)$. Analogously we write $f \in D_\alpha^-([a, b], F)$ if we have

$$\lim_{d \in D} \sum_{i=1}^{|d|} V_i[\alpha] \, \omega_i(f) = 0;$$

we have $D_\alpha^-([a, b], F) = D_{\hat{\alpha}}^-([a, b], F)$. If $\alpha(t) \equiv t$ we write simply $D([a, b], F)$.

For the numerical Riemann integral we have

$$R([a, b], \mathbb{R}) = D([a, b], \mathbb{R}) = R^-([a, b], \mathbb{R})$$
$$= D^-([a, b], \mathbb{R}) = R^\cdot([a, b], \mathbb{R})$$
$$= D^\cdot([a, b], \mathbb{R});$$

analogously we will prove that for $\alpha \in BV([a, b], \mathbb{R})$ we have

$$R_\alpha^\cdot([a, b], \mathbb{R}) = D_\alpha^\cdot([a, b], \mathbb{R}), \quad R_\alpha^-([a, b], \mathbb{R}) = D_\alpha^-([a, b], \mathbb{R}),$$

and

$$R_\alpha([a, b], \mathbb{R}) = D_\alpha([a, b], \mathbb{R})$$

DIRICHLET AND SUBSTITUTION FORMULAS

(see Corollaries 4.10˙, 4.10⁻, and 4.10) and if α is continuous these six spaces coincide (by Theorem 1.2); these facts and Theorem 4.1˙ (and the analogous results, "Theorem 4.1⁻" and "Theorem 4.1" for f and $\int$, respectively) justify the definitions given above.

THEOREM 4.1˙. Given a BT (X, F, G), $\alpha \in BV([a, b], X)$, and $f \in D_\alpha^\cdot([a, b], F)$, we have:

(a) There exists $\int_a^{b\cdot} d\alpha(t) \cdot f(t) \in G$.

(b) $\| \int_a^{b\cdot} d\alpha(t) \cdot f(t) \| \leq V[\alpha] \, \|f\|$.

Proof. (a) Given $d, \overline{d} \in D$ with $d \leq \overline{d}$, we write $I_i = [t_{i-1}, t_i]$, $\overline{I}_j = [\overline{t}_{j-1}, \overline{t}_j]$. The existence of the integral follows from

$$\| \sigma_{d,\xi} \cdot (f; \alpha) - \sigma_{\overline{d},\overline{\xi}} \cdot (f; \alpha) \|$$

$$= \left\| \sum_{i=1}^{|d|} \sum_{\overline{I}_j \subset I_i} \left(\alpha(\overline{t}_j) - \alpha(\overline{t}_{j-1}) \right) \left(f(\xi_i^\cdot) - f(\overline{\xi}_j^\cdot) \right) \right\|$$

$$\leq \sum_{i=1}^{|d|} \sum_{\overline{I}_j \subset I_i} V_j[\alpha] \, \omega_i^\cdot(f) = \sum_{i=1}^{|d|} V_i[\alpha] \, \omega_i^\cdot(f).$$

The proof of (b) is immediate.

Example. In particular there exists $\int_a^{b\cdot} d\hat{\alpha}(t) \, f(t) \in F$.

THEOREM 4.2˙. Given $\alpha \in BV([a, b], X)$, a BT (E, F, G), $f \in D_\alpha^\cdot([a, b], E)$, and $g \in D_\alpha^\cdot([a, b], F)$, we have $f \cdot g \in D_\alpha^\cdot([a, b], G)$.

__Proof__. For $d \in D$ we have $\omega_i^{\bullet}(f \cdot g) \leq \|f\| \omega_i^{\bullet}(g) + \|g\| \omega_i^{\bullet}(f)$; hence the result follows from

$$\sum_{i=1}^{|d|} V_i[\alpha] \omega_i^{\bullet}(f \cdot g) \leq \|f\| \sum_{i=1}^{|d|} V_i[\alpha] \omega_i^{\bullet}(g) + \|g\| \sum_{i=1}^{|d|} V_i[\alpha] \omega_i^{\bullet}(f).$$

__Remark__. It is immediate that for $D_{\overline{\alpha}}$ with $\smallint_a^b$ and for D_α with $\smallint_a^b$ we have theorems analogous to 4.1˙ and 4.2˙.

Given a function $\varphi: [a, b] \to \mathbb{R}_+$ and $d \in D$, we write $M_i^{\bullet}(\varphi) = \sup\{\varphi(t) \mid t \in]t_{i-1}, t_i[\}$ and $M_i(\varphi) = \sup\{\varphi(t) \mid t \in [t_{i-1}, t_i]\}$, $i = 1, 2, \ldots, |d|$. If $v: [a, b] \to \mathbb{R}$ is nondecreasing, we consider the interior upper sums

$$S_d^{\bullet}(\varphi, v) = \sum_{i=1}^{|d|} M_i^{\bullet}(\varphi) \left(v(t_i) - v(t_{i-1}) \right)$$

and the *interior upper integral*

$$\overline{\int_a^b}{}^{\bullet} \varphi(t) \, dv(t) = \inf_{d \in D} S_d^{\bullet}(\varphi, v);$$

analogously we define de upper sums

$$S_d(\varphi, v) = \sum_{i=1}^{|d|} M_i(\varphi) \left(v(t_i) - v(t_{i-1}) \right)$$

and the *upper integral*

$$\overline{\int_a^b} \varphi(t) \, dv(t) = \inf_{d \in D} S_d(\varphi, v).$$

If F is a Banach space, $f: [a, b] \to F$ a bounded function, and $\alpha \in BV([a, b], X)$, we define

$$\|f\|_{\widehat{\alpha}}^{\bullet} = \overline{\int_a^b}{}^{\bullet} \|f(t)\| \, d\widehat{\alpha}(t) \quad \text{and} \quad \|f\|_{\widehat{\alpha}} = \overline{\int_a^b} \|f(t)\| \, d\widehat{\alpha}(t);$$

$\overline{R}_{\widehat{\alpha}}^{\bullet}([a, b], F)$ $\left(\overline{R}_{\widehat{\alpha}}([a, b], F) \right)$ denotes the space of all bounded functions $f: [a, b] \to F$ endowed with the seminorm $\| \; \|_{\widehat{\alpha}}^{\bullet}$ $[\| \; \|_{\widehat{\alpha}}]$.

PROPOSITION 4.3˙ (4.3¯). $\overline{R}_{\hat{\alpha}}^{\cdot}([a, b], F)$ $[\overline{R}_{\hat{\alpha}}^{-}([a, b], F)]$ depends only on $\hat{\alpha}$ and not on α.

We write $f \in S_{\alpha}^{\cdot}([a, b], F)$ $[f \in S_{\alpha}^{-}([a, b], F)]$ if for every $\varepsilon > 0$ there exists a step function $f_{\varepsilon} \in E([a, b], F)$ such that $\|f - f_{\varepsilon}\|_{\hat{\alpha}}^{\cdot} < \varepsilon$ $[\|f - f_{\varepsilon}\|_{\hat{\alpha}}^{-} < \varepsilon]$. By Proposition 4.3˙ (4.3¯) we have

PROPOSITION 4.4˙ (4.4¯). $S_{\alpha}^{\cdot}([a, b], F)$ $[S_{\alpha}^{-}([a, b], F)]$ depends only on $\hat{\alpha}$.

If $\alpha \in BV([a, b], X)$, for every $t \in]a, b]$ there exists $\alpha(t-) = \lim_{\varepsilon \downarrow 0} \alpha(t - \varepsilon)$ and we have $\hat{\alpha}(t) - \hat{\alpha}(t-) = \|\alpha(t) - \alpha(t-)\|$; for every $t \in [a, b[$ there exists $\alpha(t+)$ and we have $\hat{\alpha}(t+) - \hat{\alpha}(t) = \|\alpha(t+) - \alpha(t)\|$ (see [2, Theorem I.2.7 and Proposition I.2.8]); since $\hat{\alpha}$ is nondecreasing we have

$$\sum_{a < t \leq b} \left(\hat{\alpha}(t) - \hat{\alpha}(t-)\right) + \sum_{a \leq t < b} \left(\hat{\alpha}(t+) - \hat{\alpha}(t)\right) \leq V[\hat{\alpha}],$$

it follows that the series $\sum_{a < t \leq b} \|\alpha(t) - \alpha(t-)\|$ and $\sum_{a \leq t < b} \|\alpha(t+) - \alpha(t)\|$ are absolutely convergent and therefore the *jump component* α_s of α is well defined:

$$\alpha_s = \sum_{a < \tau \leq b} \chi_{[\tau, b]} \left(\alpha(\tau) - \alpha(\tau-)\right) + \sum_{a \leq \tau < b} \chi_{]\tau, b]} \left(\alpha(\tau+) - \alpha(\tau)\right);$$

the *continuous component* $\alpha_c = \alpha - \alpha_s$ is a continuous function of bounded variation (α_c is continuous, since for every $t \in]a, b]$ ($t \in [a, b[$) we have $\alpha_s(t) - \alpha_s(t-) = \alpha(t) - \alpha(t-)$ $[\alpha_s(t+) - \alpha_s(t) = \alpha(t+) - \alpha(t)]$). It is immediate that $\hat{\alpha}_s = (\hat{\alpha})_s$ and $\hat{\alpha}_c = (\hat{\alpha})_c$.

If $v: [a, b] \to \mathbb{R}$ is nondecreasing, we denote by m_v the measure defined by v. (We recall that for $a < c < d < b$ we have $m_v([a, c[) = v(c-) - v(a)$, $m_v(]c, d[) = v(d-) - v(c+)$,

$m_v(]d, b]) = v(b) - v(d+)$, $m_v(\{c\}) = v(c+) - v(c-)$, etc.) Given a Banach space F and a function f: $[a, b] \to F$ we denote by D^f the set of all points of discontinuity of f.

THEOREM 4.5*. Let X and F be Banach spaces, $\alpha \in BV([a, b], X)$ and f: $[a, b] \to F$ a bounded function; the following properties are equivalent:

I*. $f \in D_\alpha^*([a, b], F)$; $\hat{I}^*$. $f \in D_{\hat{\alpha}}^*([a, b], F)$.

II*. $f \in S_\alpha^*([a, b], F)$; $\hat{II}^*$. $f \in S_{\hat{\alpha}}^*([a, b], F)$.

III*. $\lim\limits_{d \in D} \sum\limits_{i=1}^{|d|} \omega_i^*(f) \|\alpha(t_i) - \alpha(t_{i-1})\| = 0$.

IV*a. $m_v(D^f) = 0$ where $v = \hat{\alpha}_c$.

IV*b. If $\alpha(t+) \neq \alpha(t)$ then there exists $f(t+)$, and if $\alpha(t-) \neq \alpha(t)$ then there exists $f(t-)$.

Proof. I* $\iff$ $\hat{I}^*$ follows from the definition of $D_\alpha^*([a, b], F)$, and II* $\iff$ $\hat{II}^*$ follows from Proposition 4.4*; it is obvious that I* $\implies$ III*.

III* $\implies$ I*: From the definition of $V[\alpha]$ it follows that for every $\varepsilon > 0$ there exists a $d_\varepsilon \in D$ such that for $d \geq d_\varepsilon$ we have

$$V[\alpha] - \sum_{i=1}^{|d|} \|\alpha(t_i) - \alpha(t_{i-1})\| < \varepsilon;$$

hence in

$$\sum_{i=1}^{|d|} V_i[\alpha] \omega_i^*(f)$$

$$= \sum_{i=1}^{|d|} \|\alpha(t_i) - \alpha(t_{i-1})\| \omega_i^*(f)$$

$$+ \left(\sum_{i=1}^{|d|} V_i[\alpha] \omega_i^*(f) - \sum_{i=1}^{|d|} \|\alpha(t_i) - \alpha(t_{i-1})\| \omega_i^*(f) \right)$$

DIRICHLET AND SUBSTITUTION FORMULAS

$$\leq \sum_{i=1}^{|d|} \|\alpha(t_i) - \alpha(t_{i-1})\| \, \omega_i^{\bullet}(f)$$

$$+ \sum_{i=1}^{|d|} \left(V_i[\alpha] - \|\alpha(t_i) - \alpha(t_{i-1})\| \right) \omega_d^{\bullet}(f)$$

the first summand becomes arbitrarily small by the hypothesis III$^{\bullet}$ and the second one by the reasoning we have just given.

I$^{\bullet} \Longrightarrow$ II$^{\bullet}$: Given $f \in D_\alpha^{\bullet}([a, b], F)$, we take $d \in D$ such that $\Sigma_{i=1}^{|d|} V_i[\alpha] \, \omega_i^{\bullet}(f) < \varepsilon$. Then we have $f_{d,\xi^{\bullet}} \in E([a, b], F)$ and

$$\|f - f_{d,\xi^{\bullet}}\|_{\hat\alpha}^{\bullet} = \sum_{i=1}^{|d|} \int_{t_{i-1}}^{t_i} \|f(t) - f(\xi_i^{\bullet})\| \, d\hat\alpha(t)$$

$$\leq \sum_{i=1}^{|d|} V_i[\alpha] \, \omega_i^{\bullet}(f) < \varepsilon.$$

II$^{\bullet} \Longrightarrow$ I$^{\bullet}$: Given $f \in S_\alpha^{\bullet}([a, b], F)$ let $g \in E([a, b], F)$ be such that $\|f - g\|_{\hat\alpha}^{\bullet} < \varepsilon$, hence there exists $d \in D$ such that

$$\sum_{i=1}^{|d|} M_i^{\bullet}(\|f(\ ) - g(\ )\|) \left(\hat\alpha(t_i) - \hat\alpha(t_{i-1}) \right) < \varepsilon.$$

We may refine the division d and suppose that $g = \Sigma_{i=1}^{|d|} c_i \chi_{|t_{i-1}, t_i}$ and therefore we have

$$\sum_{i=1}^{|d|} M_i^{\bullet}(\|f(\ ) - c_i\|) \left(\hat\alpha(t_i) - \hat\alpha(t_{i-1}) \right) < \varepsilon.$$

But

$$\omega_i^{\bullet}(f) = \sup\{\|f(t) - f(s)\| \mid t, s \in \,]t_{i-1}, t_i[\,\}$$

$$\leq \sup\{\|f(t) - c_i\| + \|c_i - f(s)\| \mid t, s \in \,]t_{i-1}, t_i[\,\}$$

$$\leq 2 M_i^{\bullet}(\|f(\ ) - c_i\|)$$

and therefore $\Sigma_{i=1}^{|d|} V_i[\alpha] \, \omega_i^{\bullet}(f) < 2\varepsilon$.

We complete the proof of the theorem by showing that
I˙ $\iff$ IV˙. Since $\hat{\alpha} = \hat{\alpha}_c + \hat{\alpha}_s$, it is immediate that we have
I˙ if and only if we have I˙a and I˙b, where

I˙a. $\lim\limits_{d \in D} \sum\limits_{i=1}^{|d|} \omega_i^{\cdot}(f) \left[\hat{\alpha}_c(t_i) - \hat{\alpha}_c(t_{i-1})\right] = 0,$

I˙b. $\lim\limits_{d \in D} \sum\limits_{i=1}^{|d|} \omega_i^{\cdot}(f) \left[\hat{\alpha}_s(t_i) - \hat{\alpha}_s(t_{i-1})\right] = 0.$

We will prove that I˙a $\iff$ IV˙a and I˙b $\iff$ IV˙b. The equivalence of I˙a and IV˙a follows immediately from the following theorem.

THEOREM 4.6. If $v: [a, b] \to \mathbb{R}$ is continuous nondecreasing and $f: [a, b] \to F$ a bounded function, the following properties are equivalent:

(a) $\lim\limits_{d \in D} \sum\limits_{i=1}^{|d|} \omega_i^{\cdot}(f) \left[v(t_i) - v(t_{i-1})\right] = 0,$ i.e.,

$f \in D_v^{\cdot}([a, b], F).$

(b) $\lim\limits_{\Delta d \to 0} \sum\limits_{i=1}^{|d|} \omega_i(f) \left[v(t_i) - v(t_{i-1})\right] = 0,$ i.e.,

$f \in D_v([a, b], F).$

(c) $m_v(D^f) = 0.$

Proof. (a) $\implies$ (c): the set D_δ^f of points where the oscilation of f is $\geq \delta$ is closed and $D^f = \bigcup_{n \in \mathbb{N}} D_{1/n}^f$; hence if $m_v(D^f) > 0$, there exist an $n \in \mathbb{N}$ and $m > 0$ such that $m_v(D_{1/n}^f) = m$. Hence for any $d \in D$ we have

$\sum\limits_{i=1}^{|d|} \omega_i^{\cdot}(f) \left[v(t_i) - v(t_{i-1})\right]$

$\geq \sum \{\omega_i^{\cdot}(f) \left[v(t_i) - v(t_{i-1})\right] \mid \,]t_{i-1}, t_i[\, \cap D_{1/n}^f \neq \phi\} \geq \frac{m}{n}.$

DIRICHLET AND SUBSTITUTION FORMULAS

(c) $\implies$ (a): if $m_v(D^f) = 0$ then we have $m_v(D^f_{1/n}) < \varepsilon$ for all $\varepsilon > 0$ and $n \in N$. Since $D^f_{1/n}$ is compact it may be covered by a finite number of intervals $I_1, \ldots, I_p$ such that $\Sigma^p_{j=1} m_v(I_j) < \varepsilon$. At every point of $F = \complement(I_1 \cup \cdots \cup I_p)$ the oscilation of f is $< 1/n$ and since F is the union of a finite number of closed intervals, by subdivisions we may suppose that the oscilation of f in every one of these intervals is $< \varepsilon$. Hence, if d denotes the division of $[a, b]$ formed by the extremities of all these intervals we have

$$\sum_{i=1}^{|d|} \omega_i(f) \left[v(t_i) - v(t_{i-1})\right]$$

$$= \sum \{\omega_i(f) \left[v(t_i) - v(t_{i-1})\right] \mid D^f_{1/n} \cap [t_{i-1}, t_i] \neq \phi\}$$

$$+ \sum \{\omega_i(f) \left[v(t_i) - v(t_{i-1})\right] \mid D^f_{1/n} \cap [t_{i-1}, t_i] = \phi\}$$

$$\leq 2\|f\|\varepsilon + \frac{1}{n}[v(b) - v(a)];$$

hence the result, since ε and n are arbitrary.

In order to prove (a) $\implies$ (b) we need the following lemma.

LEMMA 4.7. If $d, \tilde{d} \in D$ are such that $\Delta d < \inf\{\tilde{t}_j - \tilde{t}_{j-1} \mid j = 1, \ldots, |\tilde{d}|\}$ then we have

$$\sum_{i=1}^{|d|} \omega_i(f) \left[v(t_i) - v(t_{i-1})\right]$$

$$\leq \sum_{j=1}^{|\tilde{d}|} \tilde{\omega}^*_j(f) \left[v(\tilde{t}_j) - v(\tilde{t}_{j-1})\right] + (5|\tilde{d}| - 6)\|f\| \omega_{\Delta d}(v).$$

Proof. We write $I_d = \Sigma^{|d|}_{i=1} \omega_i(f) [v(t_i) - v(t_{i-1})]$; $I_{d v \tilde{d}}$, $I^*_{\tilde{d}}$, etc., have analogous meanings. We may suppose that for $j \neq 0, |\tilde{d}|$ we have $\tilde{t}_j \neq t_i$ for $i = 1, 2, \ldots, |d| - 1$ since v is continuous and f bounded, hence we may move slightly the points $\tilde{t}_j$, $0 < j < |\tilde{d}|$, changing $I^*_{\tilde{d}}$ by an arbitrarily small amount.

We denote by $\bar{d}$ the division obtained from $\hat{d} = d \vee \tilde{d}$ by omitting the points $\hat{t}_j$ which are such that $\hat{t}_{j-1}, \hat{t}_{j+1} \notin \tilde{d}$. We denote by $I_{\bar{d}}^*$ the sum obtained by replacing $\omega_j(f)$ in $I_{\bar{d}}$ with $\omega_j^{\cdot}(f)$ if $\bar{t}_{j-1} \notin \tilde{d}$ or $\bar{t}_j \notin \tilde{d}$; then we have $I_{\bar{d}}^* \leq I_{\tilde{d}}^{\cdot}$.

We have $I_d - I_{d \vee \tilde{d}} \leq (|\tilde{d}| - 2)\|f\|\omega_{\Delta d}(v)$ because if $\hat{t}_j$ is one of the $|\tilde{d}| - 2$ points of $\tilde{d}$ that we add to d in order to obtain $\hat{d} = d \vee \tilde{d}$ then in I_d we have to replace $\omega_{[\hat{t}_{j-1}, \hat{t}_{j+1}]}(f) \times [v(\hat{t}_{j+1}) - v(\hat{t}_{j-1})]$ by

$$\omega_{[\hat{t}_{j-1}, \hat{t}_j]}(f)\left[v(\hat{t}_j) - v(\hat{t}_{j-1})\right] + \omega_{[\hat{t}_j, \hat{t}_{j+1}]}(f)\left[v(\hat{t}_{j+1}) - v(\hat{t}_j)\right];$$

hence we increase I_d by

$$c_j \leq 2\|f\| \frac{1}{2}[v(\hat{t}_{j+1}) - v(\hat{t}_{j-1})] \leq \|f\| \omega_{\Delta d}(v).$$

Hence

$$I_d \leq I_{d \vee \tilde{d}} + (|\tilde{d}| - 2)\|f\| \omega_{\Delta d}(v)$$

$$= I_{\bar{d}} + (|\tilde{d}| - 2)\|f\| \omega_{\Delta d}(v)$$

$$= [I_{\bar{d}} - I_{\bar{d}}^*] + I_{\bar{d}}^* + (|\tilde{d}| - 2)\|f\| \omega_{\Delta d}(v)$$

$$\leq 2(|\tilde{d}| - 1)2\|f\|\omega_{\Delta d}(v) + I_{\bar{d}}^* + (|\tilde{d}| - 2)\|f\| \omega_{\Delta d}(v)$$

$$\leq I_{\tilde{d}}^{\cdot} + (5|\tilde{d}| - 6)\|f\| \omega_{\Delta d}(v) \qquad Q.E.D.$$

Proof of (a) $\Longrightarrow$ (b) (of Theorem 4.6). Given $\varepsilon > 0$ we take $\tilde{d} \in D$ such that $I_{\tilde{d}}^{\cdot} \leq \varepsilon/2$, since v is uniformly continuous there exists $\delta > 0$ such that for $\Delta d \leq \delta$ we have

$$\omega_{\Delta d}(v) \leq \frac{\varepsilon}{2(5|\tilde{d}| - 6)\|f\|}$$

hence by Lemma 4.7 for $\Delta d \leq \delta$ we have

$$\sum_{i=1}^{|d|} \omega_i(f) \left[v(t_i) - v(t_{i-1})\right] \leq \varepsilon.$$

It is obvious that (b) $\Longrightarrow$ (a).

DIRICHLET AND SUBSTITUTION FORMULAS

<u>Proof of I˙b ⟺ IV˙b (of Theorem 4.5˙)</u>. Let us write $v = \hat{\alpha}_s$.

I˙b ⟹ IV˙b: For every $t \in [a, b[$ we write $\omega^\bullet_{t+}(f) = \lim_{\varepsilon \downarrow 0} \omega^\bullet_{]t, t+\varepsilon[}(f)$; hence we have $\omega^\bullet_{t+}(f) = 0$ if and only if there exists $f(t+)$. If we have $v(\bar{t}+) \neq v(\bar{t})$, then for any $d \in D$ that contains $\bar{t} = t_j$ we have

$$\sum_{i=1}^{|d|} \omega^\bullet_i(f) \left(v(t_i) - v(t_{i-1})\right) \geq \omega^\bullet_{]t_j, t_{j+1}[}(f) \left(v(t_{j+1}) - v(t_j)\right)$$
$$\geq \omega^\bullet_{\bar{t}+}(f) \left(v(\bar{t}+) - v(\bar{t})\right)$$

and analogously for the discontinuities at the left.

IV˙b ⟹ I˙b: By the definition of $v = \hat{\alpha}_s$ for every $\varepsilon > 0$ there exist $a = s_0 < s_1 < \cdots < s_n = b$ such that

$$\bar{V} = \sum_{j=1}^{n} \left(v(s_j) - v(s_j-)\right) + \sum_{j=0}^{n-1} \left(v(s_j+) - v(s_j)\right) \qquad (3)$$
$$> V[v] - \frac{\varepsilon}{6\|f\|}.$$

We take a division

$$d: a = t_0 < t_1 = a + \delta_0 < t_2 = s_1 - \delta_1 < t_3 = s_1 < t_4 = s_1 + \delta_1$$
$$< t_5 = s_2 - \delta_2 < \cdots < t_{3n} = s_n.$$

We have

$$\sum_{i=1}^{|d|} \omega^\bullet_i(f) \left(v(t_i) - v(t_{i-1})\right)$$

$$= \sum_{j=1}^{n} \omega^\bullet_{3j}(f) \left(v(t_{3j}) - v(t_{3j-1})\right)$$

$$+ \sum_{j=0}^{n-1} \omega^\bullet_{3j+1}(f) \left(v(t_{3j+1}) - v(t_{3j})\right) + \sum_{j=1}^{n} \omega^\bullet_{3j-1}(f) \left(v(t_{3j-1}) - v(t_{3j-2})\right).$$

We take $\delta_0, \delta_1, \ldots, \delta_n$ so small that we have (a), (b), (c), and (d) where

(a) $\omega_{3j}^{\bullet}(f) \leq \varepsilon/3nV[v]$ if $v(t_{3j}) - v(t_{3j}-) > 0$.

(b) $v(t_{3j}) - v(t_{3j-1}) \leq \varepsilon/6n\|f\|$ if $v(t_{3j}-) = v(t_{3j})$.

(c) $\omega_{3j+1}^{\bullet}(f) \leq \varepsilon/3nV[v]$ if $v(t_{3j}+) - v(t_{3j}) > 0$.

(d) $v(t_{3j+1}) - v(t_{3j}) \leq \varepsilon/6n\|f\|$ if $v(t_{3j}+) = v(t_{3j})$.

Then we have

$$\sum_{j=1}^{n} \omega_{3j}^{\bullet}(f) \left[v(t_{3j}) - v(t_{3j-1})\right] \leq \frac{\varepsilon}{3} \quad [\text{by (a) and (b)}],$$

$$\sum_{j=0}^{n-1} \omega_{3j+1}^{\bullet}(f) \left[v(t_{3j+1}) - v(t_{3j})\right] \leq \frac{\varepsilon}{3} \quad [\text{by (c) and (d)}],$$

$$\sum_{j=1}^{n} \omega_{3j-1}^{\bullet}(f) \left[v(t_{3j-1}) - v(t_{3j-2})\right] \leq 2\|f\| \left[V[v] - \overline{v}\right]$$

$$< \frac{\varepsilon}{3} \quad [\text{by (3)}]. \quad \text{Q.E.D.}$$

In an analogous way as Theorem 4.5· we prove

THEOREM 4.5⁻. Let X and F be Banach spaces, $\alpha \in BV([a, b], X)$ and $f: [a, b] \to F$ a bounded function; the following properties are equivalent:

I⁻. $f \in D_{\alpha}^{-}([a, b], F)$ Î⁻. $f \in D_{\hat{\alpha}}^{-}([a, b], F)$

II⁻. $f \in S_{\alpha}^{-}([a, b], F)$ Î̂I⁻. $f \in S_{\hat{\alpha}}^{-}([a, b], F)$

III⁻. $\lim_{d \in D} \sum_{i=1}^{|d|} \omega_i(f) \|\alpha(t_i) - \alpha(t_{i-1})\| = 0$

IV⁻ a. $m_v(D^f) = 0$ where $v = \hat{\alpha}_c$

b. α and f have no common left or right discontinuity, i.e., if $\alpha(t+) \neq \alpha(t)$ then there exists $f(t+) = f(t)$ and if $\alpha(t-) \neq \alpha(t)$ then there exists $f(t-) = f(t)$.

DIRICHLET AND SUBSTITUTION FORMULAS

Analogously to Theorem 4.5˙ and 4.5¯ we have

THEOREM 4.5. Let X and F be Banach spaces, $\alpha \in BV([a, b], X)$ and $f: [a, b] \to F$ a bounded function; the following properties are equivalent:

I. $f \in D_\alpha([a, b], F)$ Î. $f \in D_{\hat\alpha}([a, b], F)$

III. $\lim\limits_{\Delta d \to 0} \sum\limits_{i=1}^{|d|} \omega_i(f) \|\alpha(t_i) - \alpha(t_{i-1})\| = 0$

IVa. $m_v(D^f) = 0$ where $v = \hat\alpha_c$

b. α and f have no common points of discontinuity

V. $m_{\hat\alpha}(D^f) = 0$.

Proof. I $\iff$ Î follows from the definition of $D_\alpha([a, b], F)$; I $\implies$ III is obvious and the proof of III $\implies$ I is analogous to the proof of III˙ $\implies$ I˙ in Theorem 4.5˙. The equivalence of Î, IV, and V follows immediately from Theorem 4.6 and from

LEMMA 4.8. Under the hypothesis of Theorem 4.5 the following properties are equivalent:

Îb. $\lim\limits_{\Delta d \to 0} \sum\limits_{i=1}^{|d|} \omega_i(f) \left[\hat\alpha_s(t_i) - \hat\alpha_s(t_{i-1})\right] = 0$.

IVb. α and f have no common points of discontinuity.

Vb. $m_u(D^f) = 0$ where $u = \hat\alpha_s$.

Proof. Îb $\implies$ Vb: We have $m_u(D^f) = \Sigma_{t \in D^f} [u(t+) - u(t-)]$ [where we take $u(b+) = u(b)$ and $u(a-) = u(a)$]. Hence if we have $m_u(D^f) > 0$ there exists $\bar t \in D^f$ such that $u(\bar t+) - u(\bar t-) > 0$. Let us suppose that $\bar{t} \in]a, b[$ (if $\bar{t} = a$ or $\bar{t} = b$, the reasoning is analogous); for any $d \in D$ and $]t_{j-1}, t_j[\ni \bar{t}$ we have

$$\sum_{i=1}^{|d|} \omega_i(f) \left(u(t_i) - u(t_{i-1})\right) \geq \omega_j(f) \left(u(t_j) - u(t_{j-1})\right)$$

$$\geq \omega_{\bar{t}}(f) \left(u(\bar{t}+) - u(\bar{t}-)\right)$$

i.e., we do not have Îb. Vb $\Longrightarrow$ IVb is obvious since for every $t \in D^f$ we have

$$m_u(D^f) \geq m_u(\{t\}) = u(t+) - u(t-) = \|\alpha(t+) - \alpha(t-)\|.$$

IVb $\Longrightarrow$ Îb: If α and f have no common point of discontinuity then given $\varepsilon > 0$, for every $s \in [a, b]$ there exists $\delta_s > 0$ such that if $s \in]a, b[$ we have

$$\omega_{[s-\delta_s, s+\delta_s]}(f) \leq \frac{\varepsilon}{2V[u]} \quad \text{or} \quad u(s + \delta_s) - u(s - \delta_s) \leq \frac{\varepsilon}{2\|f\|};$$

if $s = a$ we have

$$\omega_{[a, a+\delta_a]}(f) \leq \frac{\varepsilon}{2V[u]} \quad \text{or} \quad u(a+) - u(a) \leq \frac{\varepsilon}{2\|f\|}$$

and analogously for $s = b$. Let $[a, a + \delta_a[,]s_j - \delta_j, s_j + \delta_j[,]b - \delta_b, b]$, $j = 1, \ldots, m$ (where $\delta_j = \delta_{s_j}$) be a finite subcovering of the open covering $[a, a + \delta_a[,]s - \delta_s, s + \delta_s[$ ($s \in]a, b[$), $]b - \delta_b, b]$ of $[a, b]$. If $\delta > 0$ is such that any interval $[c, d] \subset [a, b]$ with $d - c \leq \delta$ is contained in one of the intervals of the finite subcovering then for $d \in D$ with $\Delta d \leq \delta$ we have

$$\sum_{i=1}^{|d|} \omega_i(f) \left(u(t_i) - u(t_{i-1})\right) \leq \varepsilon.$$

Theorem 4.6 implies

THEOREM 4.5^c. Let X and F be Banach spaces; if $\alpha \in BV([a, b], X)$ is a continuous function we have $D_\alpha^{\bullet}([a, b], F) = D_\alpha([a, b], F)$

DIRICHLET AND SUBSTITUTION FORMULAS   173

and the following properties are equivalent: I˙, $\hat{\text{I}}$˙, II˙, $\widehat{\text{II}}$˙, III˙, I⁻, $\hat{\text{I}}$⁻, II⁻, $\widehat{\text{II}}$⁻, III⁻, I, $\hat{\text{I}}$, III, and V (of Theorems 4.5˙, 4.5⁻, and 4.5).

COROLLARY 4.9. Given a BT (E, F, G), $\alpha \in AC([a, b], E)$ (i.e., $\alpha: [a, b] \to E$ is absolutely continuous) and $f \in D([a, b], F)$ then there exists $\int_a^b d\alpha(t) \cdot f(t)$ and $f \in D_\alpha([a, b], F)$.

Proof. By Theorem 4.5 (and by Theorem 4.1) it is enough to prove that $f \in D_\alpha([a, b], F)$ or that $m_{\hat{\alpha}}(D^f) = 0$. Since $f \in D([a, b], F)$, D^f has (Lebesgue) measure zero and since α is absolutely continuous so is $\hat{\alpha}$, and therefore $m(D^f) = 0$ implies $m_{\hat{\alpha}}(D^f) = 0$.

COROLLARY 4.10˙. If $v \in BV([a, b], \mathbb{R})$ we have $R_v^\cdot([a, b], \mathbb{R}) = D_v^\cdot([a, b], \mathbb{R}) = D_{\hat{v}}^\cdot([a, b], \mathbb{R}) = R_{\hat{v}}^\cdot([a, b], \mathbb{R})$.

Proof. By Theorem 4.5˙ we have $D_v^\cdot([a, b], \mathbb{R}) \subset R_v^\cdot([a, b], \mathbb{R})$; hence it remains to prove that $R_v^\cdot([a, b], \mathbb{R}) \subset D_v^\cdot([a, b], \mathbb{R})$. If $f \in R_v^\cdot([a, b], \mathbb{R})$ then given $\varepsilon > 0$ there exists a $d_\varepsilon \in D$ such that for every $d \in D$ with $d \geq d_\varepsilon$ we have

$$|\sigma_{d,\xi}\cdot(f; v) - \sigma_{d,\eta}\cdot(f; v)| \leq \frac{\varepsilon}{2};$$

for every $i = 1, 2, \ldots, |d|$ we take $\xi_i^\cdot, \eta_i^\cdot \in \,]t_{i-1}, t_i[$ such that

$$f(\xi_i^\cdot) - f(\eta_i^\cdot) \geq \omega_i^\cdot(f) - \frac{\varepsilon}{2V[v]} \quad \text{if} \quad v(t_i) - v(t_{i-1}) \geq 0$$

and such that

$$f(\eta_i^\cdot) - f(\xi_i^\cdot) \geq \omega_i^\cdot(f) - \frac{\varepsilon}{2V[v]} \quad \text{if} \quad v(t_i) - v(t_{i-1}) < 0;$$

then we have

$$\sum_{i=1}^{|d|} \omega_i^{\bullet}(f) \left[v(t_i) - v(t_{i-1}) \right]$$

$$\leq \sum_{i=1}^{|d|} \left[f(\xi_i^{\bullet}) - f(\eta_i^{\bullet}) + \frac{\varepsilon}{2V[v]} \right] \left[v(t_i) - v(t_{i-1}) \right]$$

$$\leq |\sigma_{d,\xi} \cdot (f; v) - \sigma_{d,\eta} \cdot (f; v)| + \frac{\varepsilon}{2V[v]} \sum_{i=1}^{|d|} |v(t_i) - v(t_{i-1})|$$

$$\leq \frac{\varepsilon}{2} + \frac{\varepsilon}{2} = \varepsilon;$$

hence by the equivalence of III$^{\bullet}$ and I$^{\bullet}$ of Theorem 4.5$^{\bullet}$ we have $f \in D_v^{\bullet}([a, b], \mathbb{R})$.

In an analogous way one proves the

COROLLARY 4.10 (4.10$^-$). If $v \in BV([a, b], \mathbb{R})$ we have

$$R_v([a, b], \mathbb{R}) = D_v([a, b], \mathbb{R}) = D_{\widehat{v}}([a, b], \mathbb{R}) = R_{\widehat{v}}([a, b], \mathbb{R})$$

$$\left[R_v^-([a, b], \mathbb{R}) = D_v^-([a, b], \mathbb{R}) = D_{\widehat{v}}^-([a, b], \mathbb{R}) = R_{\widehat{v}}^-([a, b], \mathbb{R}) \right]$$

Remark. In Corollaries 4.10$^{\bullet}$, 4.10, and 4.10$^-$ we may replace $\mathbb{R}$ by a finite-dimensional Banach space F ($\simeq \mathbb{R}^n$) considering the components of the function f.

COROLLARY 4.11$^{\bullet}$. If $v \in BV([a, b], \mathbb{R})$ we have

$$f \in R_v^{\bullet}([a, b], \mathbb{R}) \implies |f| \in R_v^{\bullet}([a, b], \mathbb{R}).$$

Proof. By Corollary 4.10$^{\bullet}$ we have $f \in D_{\widehat{v}}^{\bullet}([a, b], \mathbb{R})$, hence $|f| \in D_{\widehat{v}}^{\bullet}([a, b], \mathbb{R})$ [since $\omega_i^{\bullet}(|f|) \leq \omega_i^{\bullet}(f)$], and therefore we have $|f| \in R_v^{\bullet}([a, b], \mathbb{R})$, again by Corollary 4.10$^{\bullet}$.

DIRICHLET AND SUBSTITUTION FORMULAS

Remark. It is not difficult to prove that if $v: [a, b] \to \mathbb{R}$ is such that $v \notin BV([a, b], \mathbb{R})$ then there exists $f \in R_v^{\bullet}([a, b], \mathbb{R})$ such that $|f| \notin R_v^{\bullet}([a, b], \mathbb{R})$.

COROLLARY 4.11 (4.11$^-$). If $v \in BV([a, b], \mathbb{R})$ we have

$$f \in R_v([a, b], \mathbb{R}) \implies |f| \in R_v([a, b], \mathbb{R})$$

$$\left[f \in R_v^-([a, b], \mathbb{R}) \implies |f| \in R_v^-([a, b], \mathbb{R}) \right]$$

If X and Y are Banach spaces and $\alpha \in BV([a, b], X)$, $\beta \in BV([a, b], Y)$ we write $\alpha \leq \beta$ if there exists $\lambda > 0$ such that $\lambda\hat{\beta} - \hat{\alpha}$ is nondecreasing.

PROPOSITION 4.12$^{\bullet}$ (4.12, 4.12$^-$). Given $\alpha \leq \beta$ for any Banach space F we have

$$D_\alpha^{\bullet}([a, b], F) \supset D_\beta^{\bullet}([a, b], F),$$

$$\left[D_\alpha([a, b], F) \supset D_\beta([a, b], F), \; D_\alpha^-([a, b], F) \supset D_\beta^-([a, b], F) \right].$$

Proof. The proof follows immediately from

$$V_i[\alpha] = V_i[\hat{\alpha}] \leq \lambda V_i[\hat{\beta}] = \lambda V_i[\beta].$$

Given $v \in BV([a, b], \mathbb{R})$ we denote by $v_\uparrow$ and $v_\downarrow$, respectively, its nondecreasing and nonincreasing components ($v_\uparrow = \frac{1}{2}(\hat{v} + v)$ and $v_\downarrow = \frac{1}{2}(\hat{v} - v)$).

THEOREM 4.13$^{\bullet}$. Given a Banach space F and $v \in BV([a, b], \mathbb{R})$ we have

$$D_v^{\bullet}([a, b], F) = D_{v_\uparrow}^{\bullet}([a, b], F) \cap D_{v_\downarrow}^{\bullet}([a, b], F)$$

$$= D_{\hat{v}}^{\bullet}([a, b], F).$$

Proof. Since $v_\uparrow, v_\downarrow \leq v \leq \hat{v} = v_\uparrow + v_\downarrow$ we have, by Proposition 3.12$^{\bullet}$, that

$$D_{v_\uparrow}^{\bullet}([a, b], F) \cap D_{v_\downarrow}^{\bullet}([a, b], F) \supset D_v^{\bullet}([a, b], F) = D_{\hat{v}}^{\bullet}([a, b], F);$$

from $V_i[v] = V_i[v_\uparrow] + V_i[v_\downarrow]$ follows the other inclusion.

Analogously we have

THEOREM 4.13 (4.13$^-$). Given $v \in BV([a, b], \mathbb{R})$ and a Banach space F we have

$$D_v([a, b], F) = D_{v_\uparrow}([a, b], F) \cap D_{v_\downarrow}([a, b], F) = D_{\hat{v}}([a, b], F),$$

$$\left(D_v^-([a, b], F) = D_{v_\uparrow}^-([a, b], F) \cap D_{v_\downarrow}^-([a, b], F) = D_{\hat{v}}^-([a, b], F) \right).$$

Let $v: [a, b] \to \mathbb{R}$ be a nondecreasing function; given a Banach space F and a bounded function $f: [a, b] \to F$ we write

$$f_{d,\xi} \to f \quad v\text{-a.e.} \quad (v\text{-almost everywhere})$$

if for any sequence $(d_n, \xi^{(n)}) \in \mathcal{D}$ with $\Delta d_n \to 0$ we have

$$f_{d_n, \xi^{(n)}}(t) \to f(t)$$

for all points $t \in [a, b]$ outside of a set of v-measure zero.

DIRICHLET AND SUBSTITUTION FORMULAS

THEOREM 4.14. Let X and F be Banach spaces and $\alpha \in BV([a, b], X)$; we have $f \in D_\alpha([a, b], F)$ if and only if $f_{d,\xi} \to f$ $\hat{\alpha}$-a.e.

Proof. We write $v = \hat{\alpha}$; if there exists a sequence $(d_n, \xi^{(n)}) \in \mathcal{D}$ with $\Delta d_n \to 0$ and such that $f_{d_n, \xi^{(n)}}(t) \not\to f(t)$ at all points of a set $M \subset [a, b]$ with $m_v(M) > \delta > 0$, then for every $p \in \mathbb{N}$ let M_p be the set of all points $t \in M$ such that

$$\| f_{d_n, \xi^{(n)}}(t) - f(t) \| > \frac{1}{p}$$

for an infinity of elements $n \in \mathbb{N}$. Since we have $M_p \subset M_{p+1}$, $\bigcup_{p \in \mathbb{N}} M_p = M$, and $m_v(M) > \delta$ there exists a $q \in \mathbb{N}$ such that $m_v(M_q) > \delta$, i.e., such that at every point $t \in M_q$ we have

$$\| f_{d_n, \xi^{(n)}}(t) - f(t) \| > \frac{1}{q}$$

for an infinity of elements $n \in \mathbb{N}$. If $t \in M_q$ we have

$$\omega_{[t-\varepsilon, t]}(f) + \omega_{[t, t+\varepsilon]}(f) > \frac{1}{q}$$

for all $\varepsilon > 0$, hence for any $d \in D$ the union of the intervals $[t_{i-1}, t_i]$ that contain points of M_q has v-measure $> \delta$, and therefore we have

$$2 \sum_{i=1}^{|d|} \omega_i(f) \left(v(t_i) - v(t_{i-1}) \right) > \frac{1}{q} \delta,$$

i.e., $f \notin D_v([a, b], F)$.

Reciprocally we will show that if $f: [a, b] \to F$ is a bounded function such that $f \notin D_v([a, b], F)$ then there exists a sequence $(d_n, \xi^{(n)}) \in \mathcal{D}$ with $\Delta d_n \to 0$ such that we do not have $f_{d_n, \xi^{(n)}} \to f$ v-a.e.

Indeed, since f is bounded and $f \notin D_v([a, b], F)$, by Theorem 4.5 we have $m_v(D^f) > 0$, hence there exists $\delta > 0$ such that $m_v(D_\delta^f) = m > 0$. Therefore if we take any division $d = d_n \in D$ we have

$$\sum_{i=0}^{|d|} m_v(\{t_i\} \cap D_\delta^f) + \sum_{i=1}^{|d|} m_v(]t_{i-1}, t_i[\cap D_\delta^f) \geq m;$$

if $m_v(]t_{i-1}, t_i[\cap D_\delta^f) > 0$ we have diam $f(]t_{i-1}, t_i[) \geq \delta$ and if $m_v(\{t_i\} \cap D_\delta^f) > 0$ we have diam $f(]t_{i-1}, t_i]) \geq \delta/2$ or diam $f([t_i, t_{i+1}[) \geq \delta/2$. Hence we have

(a) $\sum \{m_v(]t_{i-1}, t_i]) \mid \text{diam } f(]t_{i-1}, t_i]) \geq \frac{\delta}{2}\} \geq \frac{m}{2}$

or

(b) $\sum \{m_v([t_{i-1}, t_i[) \mid \text{diam } f([t_{i-1}, t_i[) \geq \frac{\delta}{2}\} \geq \frac{m}{2}$.

Let us consider case (a) (the reasoning is analogous in case (b)); if $m_v(]t_{i-1}, t_i] \cap D_\delta^f) = m_i > 0$ and diam $f(]t_{i-1}, t_i]) \geq \delta/2$ there exist points $\bar{s}_i, \bar{t}_i \in]t_{i-1}, t_i]$ such that $\|f(\bar{s}_i) - f(\bar{t}_i)\| > \delta/3$, and by the theorem of Hahn-Banach there exist $\varphi \in F'$ and $c \in \mathbb{R}$ such that

$$H = \varphi^{-1}(]-\infty, c]) \supset \{x \in F \mid \|x - f(\bar{s}_i)\| \leq \frac{\delta}{6}\}$$

and

$$\complement H = \varphi^{-1}(]c, \infty[) \supset \{x \in F \mid \|x - f(\bar{t}_i)\| \leq \frac{\delta}{6}\}.$$

We have

$$m_v(]t_{i-1}, t_i] \cap D_\delta^f \cap f^{-1}(H)) \geq \frac{1}{2} m_i$$

or

$$m_v(]t_{i-1}, t_i] \cap D_\delta^f \cap f^{-1}(\complement H)) \geq \frac{1}{2} m_i.$$

In the first case we take $\xi_i^{(n)} \in]t_{i-1}, t_i] \cap D_\delta^f$ such that $f(\xi_i^{(n)}) = f(\bar{t}_i)$, then we have $\|f(\xi_i^{(n)}) - f(t)\| \geq \delta/6$ for every $t \in]t_{i-1}, t_i] \cap D_\delta^f \cap f^{-1}(H)$, i.e., on a set with v-measure $\geq \frac{1}{2} m_i$; analogously we take $f(\xi_i^{(n)}) = f(\bar{s}_i)$ in the second case and we choose $\xi_i^{(n)} \in \,]t_{i-1}, t_i]$ arbitrarily if we do not have $m_v(]t_{i-1}, t_i] \cap D_\delta^f) > 0$ with diam $f(]t_{i-1}, t_i]) \geq \delta/2$. Thus we have

$$\|f_{d_n, \xi^{(n)}}(t) - f(t)\| \geq \frac{\delta}{6}$$

on a set of points $t \in [a, b]$ that [by (a)] has v-measure greater then

$$\frac{1}{2} \sum \{m_v(]t_{i-1}, t_i] \cap D_\delta^f) \mid \text{diam } f(]t_{i-1}, t_i]) \geq \frac{\delta}{2}\} \geq \frac{1}{4} m.$$

Hence by the theorem of Egoroff we cannot have $f_{d_n, \xi^{(n)}} \to f$ v-a.e.

COROLLARY 4.15. The elements of $D_\alpha([a, b], F)$ are Bochner-Lebesgue integrable (see Sec. 5) with respect to the measure defined by $\hat{\alpha}$.

Proof. With the notations of Theorem 4.14, if $f \in D_\alpha([a, b], F)$ we have $f_{d_n, \xi^{(n)}} \to f$ $\hat{\alpha}$-a.e.; hence the result follows by the theorem of dominated convergence.

5. THE SUBSTITUTION FORMULA (II)

The purpose of this section is to prove the following versions of the substitution formula:

THEOREM 5.8°. Let $(E_1, \ldots, E_{123})$ be a BAS, $\alpha \in BV([a, b], E_1)$, $f \in R_\alpha^\cdot([a, b], E_2)$, and $g \in D_\alpha^\cdot([a, b], E_3)$; then we have $f \cdot g \in R_\alpha^\cdot([a, b], E_{23})$ and

$$(S^{\bullet}) \qquad \int_a^b {}^{\bullet}d_t\left[\int_a^t {}^{\bullet}\, d\alpha(s) \cdot f(s)\right] g(t) = \int_a^b {}^{\bullet}\, d\alpha(t) \cdot f(t)\, g(t).$$

THEOREM 5.8. Let $(E_1, \ldots, E_{123})$ be a BAS, $\alpha \in BV([a, b], E_1)$, $f \in R_\alpha([a, b], E_2)$, and $g \in D_\alpha([a, b], E_3)$; then we have $f \cdot g \in R_\alpha([a, b], E_{23})$ and

$$(S) \qquad \int_a^b {}^{\bullet}d_t\left[\int_a^t {}^{\bullet}\, d\alpha(s) \cdot f(s)\right] g(t) = \int_a^b {}^{\bullet}\, d\alpha(t) \cdot f(t)\, g(t).$$

In order to prove these theorems we need several preliminary results.

THEOREM 5.1$^{\bullet}$. Let (E, F, G) be a BT, $\alpha \in BV([a, b], E)$, and $f \in R_\alpha^{\bullet}([a, b], F)$; we define $I_{f;\alpha}(t) = \int_a^t {}^{\bullet}\, d\alpha(s) \cdot f(s)$; we have

(a) $\quad I_{f;\alpha} \in BV([a, b], G)$.

(b) $\quad I_{f;\alpha} \leq \alpha$.

(c) $\quad V[I_{f;\alpha}] \leq \|f\|\, V[\alpha]$.

Proof. (a) and (c) follow from

$$\sum_{i=1}^{|d|} \|I_{f;\alpha}(t_i) - I_{f;\alpha}(t_{i-1})\| = \sum_{i=1}^{|d|} \left\|\int_{t_{i-1}}^{t_i} {}^{\bullet}\, d\alpha(s) \cdot f(s)\right\|$$

$$\leq \sum_{i=1}^{|d|} V_i[\alpha]\, \|f\|$$

$$= V[\alpha]\, \|f\|.$$

(b) follows from the fact that for $a \leq s < t \leq b$ we have

$$\|I_{f;\alpha}(t) - I_{f;\alpha}(s)\| = \left\|\int_s^t {}^{\bullet}\, d\alpha(\sigma) \cdot f(\sigma)\right\|$$

DIRICHLET AND SUBSTITUTION FORMULAS 181

$$\leq V_{[s,t]}[\alpha] \, \|f\|$$

$$= \|f\| [\hat{\alpha}(t) - \hat{\alpha}(s)].$$

In the same way we have the following:

THEOREM 5.1. Let (E, F, G) be a BT, $\alpha \in BV([a, b], E)$, and $f \in R_\alpha([a, b], F)$; we define $I_{f;\alpha}(t) = \int_a^t d\alpha(s) \cdot f(s)$; we have

(a) $I_{f;\alpha} \in BV([a, b], G)$.

(b) $I_{f;\alpha} \leq \alpha$.

(c) $V[I_{f;\alpha}] \leq \|f\| \, V[\alpha]$.

THEOREM 5.2˙. Under the hypothesis of Theorem 5.8˙ we have

(a) $g \in D^{\cdot}_{I_{f;\alpha}}([a, b], E_3)$.

(b) $\left\| \int_a^b \cdot d_t \left(\int_a^t \cdot d\alpha(s) \cdot f(s) \right) g(t) - \sum_{i=1}^{|d|} \int_{t_{i-1}}^{t_i} \cdot d\alpha(s) \, f(s) \cdot g(\xi_i^{\cdot}) \right\| \leq \sum_{i=1}^{|d|} V_i[I_{f;\alpha}] \, \omega_i^{\cdot}(g)$.

Proof. By Theorem 5.1˙ and Proposition 4.12˙ we have (a); hence the first integral in (b) exists; (b) is then immediate.

Analogously we have

THEOREM 5.2. Under the hypothesis of Theorem 5.8 we have

(a) $g \in D_{I_{f;\alpha}}([a, b], E_3)$.

(b) $\left\| \int_a^b d_t \left(\int_a^t d\alpha(s) \cdot f(s) \right) g(t) - \sum_{i=1}^{|d|} \int_{t_{i-1}}^{t_i} d\alpha(s) \, f(s) \cdot g(\xi_i) \right\| \leq \sum_{i=1}^{|d|} V_i[I_{f;\alpha}] \, \omega_i(g).$

THEOREM 5.3*. Let $(E_1, \ldots, E_{123})$ be a BAS, $\alpha \in BV([a, b], E_1)$, $f \in D_\alpha^*([a, b], E_2)$, and $g: [a, b] \to E_3$; if one of the integrals in (S*) exists, so does the other and both are equal.

Proof. Given $(d, \xi^*) \in \mathcal{D}^*$ we consider the approximating sums to both integrals; we have

$\left\| \sum_{i=1}^{|d|} \int_{t_{i-1}}^{t_i} d\alpha(s) \, f(s) \cdot g(\xi_i^*) \right.$

$\left. - \sum_{i=1}^{|d|} \left[\alpha(t_i) - \alpha(t_{i-1}) \right] \cdot f(\xi_i^*) \, g(\xi_i^*) \right\|$

$= \left\| \sum_{i=1}^{|d|} \int_{t_{i-1}}^{t_i} d\alpha(s) \cdot \left[f(s) - f(\xi_i^*) \right] g(\xi_i^*) \right\|$

$\leq \|g\| \sum_{i=1}^{|d|} V_i[\alpha] \, \omega_i^*(f),$

and hence the result because $f \in D_\alpha^*([a, b], E_2)$.

In the same way one proves

THEOREM 5.3. Let $(E_1, \ldots, E_{123})$ be a BAS, $\alpha \in BV([a, b], E_1)$, $f \in D_\alpha([a, b], E_2)$, and $g: [a, b] \to E_3$; if one of the integrals in (S) exists, so does the other and both are equal.

LEMMA 5.4*. With the notations of Theorem 5.3*, if $g \in R_{I_{f;\alpha}}^*([a, b], E_3)$, then we have $f \cdot g \in R_\alpha^*([a, b], E_{23})$ and (S*).

DIRICHLET AND SUBSTITUTION FORMULAS 183

   Proof. By the hypothesis, the first integral of (S˙) exists; the result follows from Theorem 5.3˙.

LEMMA 5.4. With the notations of Theorem 5.3, if $g \in R_{I_{f;\alpha}}([a, b], E_3)$, then we have $f \cdot g \in R_\alpha([a, b], E_{23})$ and (S).

LEMMA 5.5˙. With the notations of Theorem 5.3˙, if $g \in D_\alpha^\cdot([a, b], E_3)$, then $g \in D_{I_{f;\alpha}}^\cdot([a, b], E_3)$ and we have (S˙).

   Proof. By Theorem 4.2˙ we have $f \cdot g \in D_\alpha^\cdot([a, b], E_{23})$ and in Theorem 5.1˙ we saw that $I_{f;\alpha} \leq \alpha$; hence by Proposition 4.12˙ we have $g \in D_{I_{f;\alpha}}^\cdot([a, b], E_3)$ and so by Lemma 5.4˙ the second integral exists too and we have (S˙).

LEMMA 5.5. With the notations of Theorem 5.3, if $g \in D_\alpha([a, b], E_3)$, then we have $g \in D_{I_{f;\alpha}}([a, b], E_3)$ and (S).

PROPOSITION 5.6˙. Under the hypothesis of Theorem 5.8˙ we have

$$\int_a^b d_t\left(\int_a^t d\alpha(s) \cdot f(s)\right) g(t) = \lim_{d \in D} \int_a^b d_t\left(\int_a^t d\alpha(s) \cdot f_{d,\xi^\cdot}(s)\right) g(t).$$

   Proof. By Theorem 5.2˙ the first integral exists. We write

$$I(t) = \int_a^t d\alpha(s) \cdot f(s) \quad \text{and} \quad I_{(d,\xi^\cdot)}(t) = \int_a^t d\alpha(s) \cdot f_{d,\xi^\cdot}(s).$$

For $(\bar{d}, \bar{\xi}^\cdot) \in \mathcal{D}^\cdot$ we have

$$\left\|\int_a^b d_t\left(\int_a^t d\alpha(s) \cdot f(s)\right) g(t) - \int_a^b d_t\left(\int_a^t d\alpha(s) \cdot f_{d,\xi^\cdot}(s)\right) g(t)\right\|$$

$$= \left\|\int_a^b dI(t) \cdot g(t) - \int_a^b dI_{(d,\xi^\cdot)}(t) \cdot g(t)\right\|$$

$$\leq \left\| \int_a^b dI_{(d,\xi\cdot)}(t) \cdot g(t) - \int_a^b d\alpha(t) \cdot f_{d,\xi\cdot}(t) \, g(t) \right\|$$

$$+ \left\| \int_a^b d\alpha(t) \cdot f_{d,\xi\cdot}(t) \, g(t) - \sum_{j=1}^{|\overline{d}|} \int_{\overline{t}_{j-1}}^{\overline{t}_j} d\alpha(t) \cdot f_{d,\xi\cdot}(t) \, g(\overline{\xi}_j^{\cdot}) \right\|$$

$$+ \left\| \sum_{j=1}^{|\overline{d}|} \int_{\overline{t}_{j-1}}^{\overline{t}_j} d\alpha(t) \cdot f_{d,\xi\cdot}(t) \, g(\overline{\xi}_j^{\cdot}) \right.$$

$$\left. - \sum_{j=1}^{|\overline{d}|} \left[I_{(d,\xi\cdot)}(\overline{t}_j) - I_{(d,\xi\cdot)}(\overline{t}_{j-1}) \right] g(\overline{\xi}_j^{\cdot}) \right\|$$

$$+ \left\| \sum_{j=1}^{|\overline{d}|} \left[(I_{(d,\xi\cdot)}(\overline{t}_j) - I(\overline{t}_j)) - (I_{(d,\xi\cdot)}(\overline{t}_{j-1}) - I(\overline{t}_{j-1})) \right] g(\overline{\xi}_j^{\cdot}) \right\|$$

$$+ \left\| \sum_{j=1}^{|\overline{d}|} \left[I(\overline{t}_j) - I(\overline{t}_{j-1}) \right] g(\overline{\xi}_j^{\cdot}) - \int_a^b dI(t) \cdot g(t) \right\|.$$

By the definition of $I_{(d,\xi\cdot)}$ the third summand is zero; by Lemma 5.5˙ the first summand is zero because $f_{d,\xi\cdot} \in E([a, b], E_2) \subset S_\alpha^\cdot([a, b], E_2) = D_\alpha^\cdot([a, b], E_2)$. The second summand is $\leq \|f\| \sum_{j=1}^{|\overline{d}|} V_j[\alpha] \, \omega_j^\cdot(g)$ for all $(d, \xi\cdot) \in \mathcal{D}^\cdot$, and so is the fifth summand by Theorem 5.2˙ and by Theorem 5.1˙ (c). Hence, given $\varepsilon > 0$, there is a $d_\varepsilon \in D$ such that for $\overline{d} \geq d_\varepsilon$ the second and fifth summands are $< \varepsilon/3$ for all $d \in D$, we fix such a $\overline{d}$; then by Proposition 1.3˙ the fourth summand becomes $< \varepsilon/3$ for all d sufficiently "large."

PROPOSITION 5.6. Under the hypothesis of Theorem 5.8 we have

$$\int_a^b d_t \left(\int_a^t d\alpha(s) \cdot f(s) \right) g(t) = \lim_{\Delta d \to 0} \int_a^b d_t \left(\int_a^t d\alpha(s) \cdot f_{d,\xi}(s) \right) g(t).$$

__Proof.__ The proof follows the steps of the proof of Proposition 5.6˙.

LEMMA 5.7*. Under the hypothesis of Theorem 5.8* we have

$$\left\| \int_a^b \bullet \, d\alpha(t) \bullet f_{d,\xi}*(t) \, g(t) - \sum_{i=1}^{|d|} \left[\alpha(t_i) - \alpha(t_{i-1}) \right] \bullet f(\xi_i^*) \, g(\xi_i^*) \right\|$$

$$\leq \|f\| \sum_{i=1}^{|d|} V_i[\alpha] \, \omega_i^*(g).$$

The proof is obvious.

LEMMA 5.7. Under the hypothesis of Theorem 5.8 we have

$$\left\| \int_a^b \bullet \, d\alpha(t) \bullet f_{d,\xi}(t) \, g(t) - \sum_{i=1}^{|d|} \left[\alpha(t_i) - \alpha(t_{i-1}) \right] \bullet f(\xi_i) \, g(\xi_i) \right\|$$

$$\leq \|f\| \sum_{i=1}^{|d|} V_i[\alpha] \, \omega_i(g).$$

Proof of Theorem 5.8*. The first integral in (S*) exists by Theorem 5.2* and is, by Proposition 5.6*, equal to

$$\lim_{d \in D} \int_a^b \bullet \, d_t \left(\int_a^t \bullet \, d\alpha(s) \bullet f_{d,\xi}*(s) \right) g(t),$$

which by Lemma 5.5* is equal to

$$\lim_{d \in D} \int_a^b \bullet \, d\alpha(t) \bullet f_{d,\xi}*(t) \, g(t);$$

by Lemma 5.7* this limit is equal to

$$\lim_{d \in D} \sum_{i=1}^{|d|} \left[\alpha(t_i) - \alpha(t_{i-1}) \right] \bullet f(\xi_i^*) \, g(\xi_i^*)$$

and by definition this is $\int_a^b \bullet \, d\alpha(t) \bullet f(t) \, g(t)$.

Proof of Theorem 5.8. The proof follows the steps of the proof of Theorem 5.8* (applying Theorem 5.2, Proposition 5.6, etc. instead of Theorem 5.2*, Proposition 5.6*, etc.).

COROLLARY 5.9*. Let (E, F, G) be a BT, $v \in BV([a, b], \mathbb{R})$, $f \in R_v^*([a, b], E)$, and $g \in D_v^*([a, b], F)$; we have

$$\int_a^b d_t \left(\int_a^t {}^\bullet f(s) \, dv(s) \right) g(t) = \int_a^b {}^\bullet f(t) \, g(t) \, dv(t).$$

COROLLARY 5.9. Let (E, F, G) be a BT, $v \in BV([a, b], \mathbb{R})$, $f \in R_v([a, b], E)$, and $g \in D_v([a, b], F)$; we have

$$\int_a^b d_t \left(\int_a^t f(s) \, dv(s) \right) g(t) = \int_a^b f(t) \, g(t) \, dv(t).$$

COROLLARY 5.10*. Let (E, F, G) be a BT, $v \in BV([a, b], \mathbb{R})$, $f \in R_v^*([a, b], E)$, and $g \in D_v^*([a, b], F)$; we have

$$\int_a^b {}^\bullet f(t) \, d_t \left(\int_a^t {}^\bullet g(s) \, dv(s) \right) = \int_a^b {}^\bullet f(t) \, g(t) \, dv(t).$$

Proof. By Theorem 5.8* the second member exists, and by Theorem 5.3* it follows that the first one exists and that they are equal.

COROLLARY 5.10. Let (E, F, G) be a BT, $v \in BV([a, b], \mathbb{R})$, $f \in R_v([a, b], E)$, and $g \in D_v([a, b], F)$; we have

$$\int_a^b f(t) \, d_t \left(\int_a^t g(s) \, dv(s) \right) = \int_a^b f(t) \, g(t) \, dv(t).$$

Remark. Even in the particular case of Corollary 5.9, when we take $v(t) \equiv t$ the proof of $\int_a^b d_t [\int_a^t f(s) \, ds] g(t) = \int_a^b f(t) g(t) \, dt$ does not become essentially simpler then the proof of the general Theorem 5.8*!

Let E be a Banach space and $1 \leq p \leq \infty$; $L_p([a, b], E)$ denotes the Banach space of (equivalence classes of) functions f: $[a, b] \to E$ that are p-integrable in the sense of Bochner and Lebesgue (see [8]). It is easy to prove that $S^-([a, b], E) \subset L_1([a, b], E)$, hence by Theorem 4.5^c we have $D([a, b], E) \subset L_1([a, b], E)$ and even $D([a, b], E) \subset L_\infty([a, b], E)$ (see also Corollary 4.15). We write $\alpha \in L_1^{(1)}([a, b], E)$ if there exists $\beta \in L_1([a, b], E)$ such that

$$\alpha(t) = \alpha(a) + {\scriptstyle L}\!\!\int_a^t \beta(s) \, ds$$

(where ${\scriptstyle L}\!\!\int$ denotes the integral of Bochner-Lebesgue), then α is absolutely continuous, differentiable a.e., and satisfies $\alpha' = \beta$ a.e. (see [8]); we have also

$$V[\alpha] = \|\alpha'\|_1 = {\scriptstyle L}\!\!\int_a^b \|\alpha'(t)\| \, dt.$$

Indeed: given $d \in D$ we have

$$\|\alpha(t_i) - \alpha(t_{i-1})\| = \left\| {\scriptstyle L}\!\!\int_{t_{i-1}}^{t_i} \alpha'(t) \, dt \right\| \leq {\scriptstyle L}\!\!\int_{t_{i-1}}^{t_i} \|\alpha'(t)\| \, dt,$$

hence $V[\alpha] \leq {\scriptstyle L}\!\!\int_a^b \|\alpha'(t)\| \, dt$. Reciprocally, it is immediate that we have $\|\alpha'(t)\| \leq \hat{\alpha}'(t)$ a.e., hence

$${\scriptstyle L}\!\!\int_a^b \|\alpha'(t)\| \, dt \leq \int_a^b \hat{\alpha}'(t) \, dt = \hat{\alpha}'(b) = V[\alpha].$$

COROLLARY 5.11. Let (E, F, G) be a BT, $\alpha \in L_1^{(1)}([a, b], E)$, and $f \in R([a, b], F)$; then there exists $\int_a^b d\alpha(t) \cdot f(t)$.

Proof. Let us take a sequence $\beta_n \in C([a, b], E)$ such that $\|\beta_n - \alpha'\|_1 \to 0$ and define $\alpha_n(t) = \alpha(a) + \int_a^t \beta_n(s) \, ds$, $t \in [a, b]$. Given $d, \bar{d} \in D$ with $\bar{d} \geq d$ we write $\bar{I}_j = [\bar{t}_{j-1}, \bar{t}_j]$ and $I_i = [t_{i-1}, t_i]$ and we have

$$\|\sigma_{\bar{d}, \bar{\xi}}(f; \alpha) - \sigma_{d, \xi}(f; \alpha)\| = \left\| \sum_{i=1}^{|d|} \sum_{\bar{I}_j \subset I_i} \left(\alpha(\bar{t}_j) - \alpha(\bar{t}_{j-1})\right) \right.$$
$$\left. \times \left(f(\bar{\xi}_j) - f(\xi_i)\right) \right\|$$

$$\leq \left\| \sum_{i=1}^{|d|} \sum_{\overline{I}_j \subset I_i} \left\{ \left(\alpha(\overline{t}_j) - \alpha_n(\overline{t}_j)\right) - \left(\alpha(\overline{t}_{j-1}) - \alpha_n(\overline{t}_{j-1})\right) \right\} \right.$$
$$\left. \times \left[f(\overline{\xi}_j) - f(\xi_i) \right] \right\|$$
$$+ \left\| \sum_{i=1}^{|d|} \sum_{\overline{I}_j \subset I_i} \left(\alpha_n(\overline{t}_j) - \alpha_n(\overline{t}_{j-1})\right) \left[f(\overline{\xi}_j) - f(\xi_i) \right] \right\|$$
$$\leq V[\alpha - \alpha_n] \, 2\|f\| + \|\sigma_{\overline{d},\overline{\xi}}(f;\alpha_n) - \sigma_{d,\xi}(f;\alpha_n)\| \, ;$$

since $V[\alpha - \alpha_n] = \|\alpha' - \alpha_n'\|_1 = \|\alpha' - \beta_n\|_1$, given $\varepsilon > 0$ there exists an $n_\varepsilon \in \mathbb{N}$ such that for $n \geq n_\varepsilon$ we have $\|\alpha' - \beta_n\|_1 \leq \varepsilon/2$, and if we fix such an n by Corollary 5.9 there exists $\int_a^b d\alpha_n(t) \cdot f(t) = \int_a^b \alpha_n'(t) f(t) \, dt$. Hence there exists $\delta > 0$ such that for $\Delta d \leq \delta$ we have

$$\|\sigma_{\overline{d},\overline{\xi}}(f;\alpha_n) - \sigma_{d,\xi}(f;\alpha_n)\| \leq \frac{\varepsilon}{2}. \quad \text{Q.E.D.}$$

COROLLARY 5.12. Let (E, F, G) be a BT, $\alpha \in L_1^{(1)}([a,b], E)$, and $f \in D([a,b], F)$; then there exists $\int_a^b \alpha'(t) f(t) \, dt = \int_a^b d\alpha(t) \cdot f(t)$.

Proof. Since $D([a,b], F) \subset L_\infty([a,b], F)$, the first integral exists. For every $d \in D$ we have

$$\left\| \int_a^b \alpha'(t) f(t) \, dt - \sum_{i=1}^{|d|} \left(\alpha(t_i) - \alpha(t_{i-1})\right) f(\xi_i) \right\|$$
$$= \left\| \sum_{i=1}^{|d|} \int_{t_{i-1}}^{t_i} \alpha'(t) \left(f(t) - f(\xi_i)\right) dt \right\|$$
$$\leq \sum_{i=1}^{|d|} \int_{t_{i-1}}^{t_i} \|\alpha'(t)\| \|f(t) - f(\xi_i)\| \, dt$$
$$\leq \sum_{i=1}^{|d|} V_i[\alpha] \, \omega_i(f)$$

that goes to zero when $\Delta d \to 0$ since by Corollary 4.9 we have $f \in D_\alpha([a,b], F)$.

REFERENCES

1. C. S. Hönig, The Green function of a linear differential equation with a lateral condition, *Bull. Amer. Math. Soc.*, 79 (1973), 587-593.
2. C. S. Hönig, The abstract Riemann-Stieltjes integral and its applications to linear differential equations with generalized boundary conditions, *Notas do Instituto de Matemática e Estatística da Universidade de São Paulo*, Série Matemática No. 1, 1973.
3. C. S. Hönig, Volterra Stieltjes integral equations with linear constraints and discontinuous solutions, *Bull. Amer. Math. Soc.*, 81 (1975), 593-598.
4. C. S. Hönig, *Volterra Stieltjes-integral equations*, Mathematics Studies 16, North-Holland Publishing Comp., Amsterdam, 1975.
5. M. Gowurin, Über die Stieltjes Integration abstrakter Funktionen, *Fund. Math.*, 27 (1936), 254-268.
6. N. Dinculeanu, *Vector Measures*, Pergamon Press, Oxford, 1967.
7. T. H. Hildebrandt, *Introduction to the Theory of Integration*, Academic Press, New York, 1963.
8. S. Bochner, Integration von Funktionen, deren Werte die Elemente eines Vektorraumes sind, *Fund. Math.*, 20 (1933), 262-276.

FUNCTIONAL DIFFERENTIAL EQUATIONS

A. F. Izé

Instituto de Ciências Matemáticas de São Carlos
Departamento de Matemática
Universidade de São Paulo
São Carlos, São Paulo, Brazil

and

O. F. Lopes

Instituto de Matemática, Estatística e Ciências de Computação
Universidade Estadual de Campinas
Campinas, São Paulo, Brazil

1. PHYSICAL AND BIOLOGICAL MOTIVATION

Many mathematical models arising in biology and physics are described by ordinary differential equations, since we assume that the transmission time of the effect is zero ou approximately zero. When the transmission time is not zero, more realistic models give rise to retarded differential equations. More general equations (of the neutral type) arise as an alternative form in the study of boundary-value problems for certain hyperbolic partial differential equations.

In what follows we give two examples of phenomena described by such equations. Some other examples can be found in the references.

1.1. Propagation of Infectious Diseases

Consider a disease (measles, for example) which presents the following cycle [1]:

birth → susceptible → exposed → infectious → immune.

The last step means that the individual does not contract the disease twice. At each instant t, the population can be divided in the following disjoint groups:

  $S(t)$ = number of susceptible individuals,

  $E(t)$ = number of exposed individuals
       (they will become infectious),

  $I(t)$ = number of infectious individuals,

  $R(t)$ = number of immune individuals.

The disease spreads as a result of the contact between infectious and susceptible individuals. Let us assume that the number of contacts depends linearly of S and I, that is, the rate of new infectious is $\beta(t) S(t) I(t)$, where $\beta(t)$ is a one-year periodic constant which is seasonally dependent. If γ is the rate at which individuals enter the population then

$$\frac{dS}{dt} = -\beta(t) S(t) I(t) + \gamma.$$

Let us assume now a deterministic hypothesis: If one individual is exposed at instant t (measured in days) he will be infected in the interval of time $[t + 12, t + 14]$. Under these hypothesis we see that the number of infected individuals at the instant

FUNCTIONAL DIFFERENTIAL EQUATIONS

t is equal to the number of exposed individuals in the interval of time $[t - 14, t - 12]$, that is,

$$I(t) = \int_{t-14}^{t-12} \beta(x) S(x) I(x) \, dx = \int_{t-14}^{t-12} [\gamma - S'(x)] \, dx$$

$$= 2\gamma + S(t - 14) - S(t - 12);$$

then

$$S'(t) = -\beta(t) S(t) [2\gamma + S(t - 14) - S(t - 12)] + \gamma,$$

which is a retarded differential equation.

1.2. Transmission Line

Consider a transmission line without loss, as represented in Fig. 1 [2, 3]:

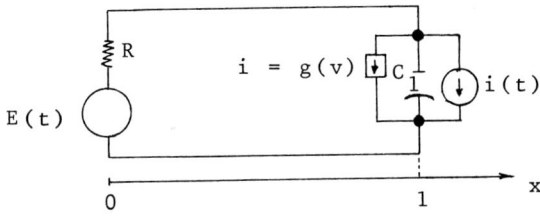

FIG. 1.

Let $v(x, t)$ and $i(x, t)$ be the voltage and the current in the line in the point x and at instant t; then v and i satisfy

$$L \frac{\partial i}{\partial t} = - \frac{\partial v}{\partial x}, \quad C \frac{\partial v}{\partial t} = -\frac{\partial i}{\partial x},$$

where L and C are the inductance and the specific capacitance of the line, respectively. The boundary conditions are

(B1) $\qquad E(t) = v(0, t) + R\, i(0, t),$

(B2) $\qquad i(1, t) = C_1 \dfrac{\partial v(1, t)}{t} + g(v(1, t)) + i(t).$

The general solution of the system is

(A) $\qquad v(x, t) = \dfrac{1}{2}\left[\phi(x - st) + \psi(x + st)\right],$

$\qquad\qquad i(x, t) = \dfrac{1}{2z}\left[\phi(x - st) - \psi(x + st)\right],$

where $s = (LC)^{-1/2}$, $z = (L/C)^{1/2}$, $Cz = 1/s$.

From (A) we can write

$$v\!\left(0, t - \tfrac{1}{s}\right) + z\, i\!\left(0, t - \tfrac{1}{s}\right) = v(1, t) + z\, i(1, t),$$

$$v\!\left(0, t - \tfrac{1}{s}\right) - z\, i\!\left(0, t - \tfrac{1}{s}\right) = v\!\left(1, t - \tfrac{2}{s}\right) - z\, i\!\left(1, t - \tfrac{2}{s}\right);$$

adding and subtracting the above equalities we have

$$2\, v\!\left(0, t - \tfrac{1}{s}\right) = v(1, t) + v\!\left(1, t - \tfrac{2}{s}\right) + z\!\left[i(1, t) - i\!\left(1, t - \tfrac{2}{s}\right)\right],$$

$$2z\, i\!\left(0, t - \tfrac{1}{s}\right) = v(1, t) - v\!\left(1, t - \tfrac{2}{s}\right) + z\!\left[i(1, t) + i\!\left(1, t - \tfrac{2}{s}\right)\right].$$

From condition (B1), at instant $t - (1/s)$ we have

$$E\!\left(t - \tfrac{1}{s}\right) = v\!\left(0, t - \tfrac{1}{s}\right) + R\, i\!\left(0, t - \tfrac{1}{s}\right)$$

and by substitution of $v(0, t - (1/s))$ and $i(0, t - (1/s))$ by the above equation, we have

$$E\!\left(t - \tfrac{1}{s}\right) = \tfrac{1}{2}\!\left(1 + \tfrac{R}{z}\right) v(1, t) + \tfrac{1}{2}\!\left(1 - \tfrac{R}{z}\right) v\!\left(1, t - \tfrac{2}{s}\right)$$

$$+ \tfrac{1}{2}\, i(1, t)(R + z) + \tfrac{1}{2}\, i\!\left(1, t - \tfrac{2}{s}\right)(R - z),$$

FUNCTIONAL DIFFERENTIAL EQUATIONS

that is,

$$i(1, t) + K\, i\!\left(1, t - \frac{2}{s}\right) = \alpha(t) - \frac{1}{z} v(1, t) + \frac{K}{z} v\!\left(1, t - \frac{2}{s}\right),$$

where

$$K = \frac{R - z}{R + z} \quad \text{and} \quad \alpha(t) = \frac{2E}{R + z}\left(t - \frac{1}{s}\right).$$

Using (B2) in the above equation and writing $v(t) = v(1, t)$, we have

$$C_1 \frac{d}{dt} v(t) + g(v(t)) + i(t)$$
$$+ K\!\left[C_1 \frac{d}{dt} v(t - \tfrac{2}{s}) + g\!\left(v(t - \tfrac{2}{s})\right) + i(t - \tfrac{2}{s})\right]$$
$$= \alpha(t) - \frac{1}{z} v(t) + \frac{K}{z} v(t - \tfrac{2}{s}),$$

or

$$C_1 \frac{d}{dt}\left[v(t) + K\, v(t - \tfrac{2}{s})\right]$$
$$= \frac{2}{R + z} E(t - \tfrac{1}{s}) - i(t) - K\, i(t - \tfrac{2}{s}) - \frac{1}{z} v(t) - g(v(t))$$
$$+ \frac{K}{z} v(t - \tfrac{2}{s}) - K\, g\!\left(v(t - \tfrac{2}{s})\right),$$

that is,

$$\frac{d}{dt}\left[v(t) + K\, v(t - h)\right]$$
$$= -\frac{1}{zC_1} v(t) - \frac{g(v(t))}{C_1} + \frac{K}{zC_1} v(t - h) - \frac{K}{C_1} g(v(t - h)) + p(t),$$

which is a functional differential equation of the neutral type.

2. GEOMETRIC THEORY FOR FUNCTIONAL DIFFERENTIAL EQUATIONS OF NEUTRAL TYPE

Consider the vector space $\mathbb{R}^n$ with norm $|\cdot|$ and let $r \geq 0$ be a real number. We note by $C([a, b], \mathbb{R}^n)$ the Banach space of continuous functions ϕ mapping the interval $[a, b]$ into $\mathbb{R}^n$, with norm $\|\phi\| = \sup|\phi(\theta)|$, $-r \leq \theta \leq 0$. Let $C = C([-r, 0], \mathbb{R}^n)$. If σ and $A \geq 0$ are real numbers, then, for every $t \in [\sigma, \sigma + A]$ and $x \in C([\sigma - r, \sigma + A], \mathbb{R}^n)$ we let x_t be a function of C defined by $x_t(\theta) = x(t + \theta)$, $-r \leq \theta \leq 0$. Geometrically the above definition means the following: for $t \in [\sigma, \sigma + A]$, consider the graph of the restriction of the function x to the interval $[t - r, t]$ shifted in a direction parallel to the t axis in order to be the graph of a function defined in the interval $[-r, 0]$; this function is x_t.

DEFINITION. If Ω is an open set of $\mathbb{R} \times \mathbb{C}$ and $f, D: \Omega \to E^n$ are continuous functions, then the relation

$$\frac{d}{dt} D(t, x_t) = f(t, x_t)$$

is a *functional differential equation*. A function $x \in C([\sigma - r, \sigma + A], \mathbb{R}^n)$ for some (σ, A), $A \geq 0$, is a solution of this equation if $(t, x_t) \in \Omega$ for $t \in [\sigma, \sigma + A]$, $D(t, x_t)$ is differentiable and satisfies the equation for $t \in [\sigma, \sigma + A]$.

For a given pair $(\sigma, \phi) \in \Omega$, we say that x is a solution through (σ, ϕ) if $x_\sigma = \phi$.

To avoid technical details, let us restrict our discussion to the case in which D is linear in ϕ. In what follows we define a class of operators D for which the initial-values problem for a functional differential equation has solutions depending continuously on the initial conditions.

Suppose that $D: \mathbb{R} \times \mathbb{C} \to E^n$ is continuous and linear in the second variable. By the Riesz representation theorem, there is

FUNCTIONAL DIFFERENTIAL EQUATIONS

a matrix $\mu(t, \theta)$, $\theta \in [-r, 0]$ such that $D(t, \phi) = \int_{-r}^{0} d\mu(t, \theta)\phi(\theta)$, $\forall \phi \in C$, where, for each t, $\mu(t, \cdot)$ is a matrix whose elements are functions of bounded variation.

DEFINITION. D is atomic at $\beta \in [-r, 0]$ if the matrix $A(t, \beta) = \mu(t, \beta_+) - \mu(t, \beta_-)$ is invertible for each t and there exists a function $\gamma(s, t)$, $s \geq 0$, with $\gamma(0, t) = 0$, $\forall t$, γ continuous in $(0, t)$ such that

$$\left| \int_{\beta-s}^{\beta+s} d\mu(t, \theta)\phi(\theta) - A(t, \beta)\phi(\theta) \right| \leq \gamma(s, t) |\phi|, \quad \forall \phi \in C.$$

Examples.

1. The operator $D\phi = \phi(0)$ is atomic in $\beta = 0$ and not atomic in $\beta = -r$. For this operator we have $Dx_t = x(t)$, and the differential equation becomes

$$\frac{d}{dt} x(t) = f(t, x_t).$$

This equation is known as a *retarded functional differential equation*.

2. The operator $D\phi = \phi(0) - B \phi(-r)$, $r > 0$ is atomic at $\beta = 0$. If B is invertible, D is also atomic at $\beta = -r$.

3. $$D\phi = \sum_{i=1}^{\infty} \frac{1}{2^i} \phi\left(-\frac{1}{2^i}\right)$$

is not atomic at $\beta = 0$.

DEFINITION. If D is atomic at $\beta = 0$, we say that the functional differential equation

$$\frac{d}{dt} D(t, x_t) = f(t, x_t)$$

is a *differential equation of the neutral type*.

Almost all theory of initial-value problems for ordinary differential equations can be extended for equations of the neutral type. Such an extension is possible using the following fixed-point theorem.

THEOREM (Krasnoselskii) [4]. Suppose that Γ is a closed, convex, and bounded set of a Banach space X. If $S: \Gamma \to X$ is a contraction, $U: \Gamma \to X$ is compact and $S(\Gamma) + U(\Gamma) = \{S(x) + U(y) \mid (x, y) \in \Gamma \times \Gamma\}$ is in Γ, then $T = S + U$ has a fixed point.

The next theorem tells us that, if S and U are continuously dependent on certain parameters, then the fixed point of $S + U$ (if it is unique) also depends continuously on these parameters.

THEOREM. Let S_k, U_k, and T_k be sequences of functions satisfying the hypothesis of the above theorem, as well as S, U, and T. Suppose that the contraction constant of S_k does not depend on k and that $\bigcup_{k \geq 1} U_k(\Gamma)$ has a compact closure. Suppose that S_k converge to S, U_k converge to U, uniformly on compact sets and that $T = S + U$ has a unique fixed point x_0. Then, for every sequence x_k of fixed points of $T_k = S_k + U_k$, x_k converge to x_0.

Using the foregoing results we can prove the following [5].

EXISTENCE THEOREM. If

$$\frac{d}{dt} D(t, x_t) = f(t, x_t)$$

is a neutral functional differential equation, then, for every $(\sigma, \phi) \in \Omega$, there exists one solution through (σ, ϕ) defined in the interval $[\sigma, \sigma + A]$, $A > 0$.

FUNCTIONAL DIFFERENTIAL EQUATIONS

UNIQUENESS THEOREM. If f is Lipschitzian on compact sets and x, y are solutions through (σ, ϕ), then $x(t) = y(t)$ for $t \geq \sigma$.

Remark. In general there is no unique "backward" solution. For example, if x is a solution of

$$\frac{dx}{dt} = \alpha\, x(t) \left[1 - x(t-1)\right]$$

through (σ, ϕ) and $\phi(0) = 0$, then $x(t) = 0$ for $t \geq \alpha$ [6].

CONTINUOUS DEPENDENCE THEOREM. Consider a sequence of neutral equations

(I_k) $\qquad\qquad \frac{d}{dt} D_k(t)\, x_t = f_k(t, x_t),$

a neutral equation

(I) $\qquad\qquad \frac{d}{dt} D(t, x_t) = f(t, x_t),$

$$D, D_k, f, f_k: \Omega \subset \mathbb{R} \times \mathbb{C} \to \mathbb{R}^n,$$

a sequence (σ_k, ϕ_k) of elements of Ω, and an element (σ, ϕ) in Ω. Consider the following hypotheses:

(i) $f_k \to f$, $D_k \to D$ uniformly on compact sets.

(ii) $\text{Var}_{[-s,0]}\, \mu_k(t, \cdot) \to 0$ when $s \to 0$ and $k \to \infty$, uniformly for t on compacts sets.

(iii) $(\sigma_k, \phi_k) \to (\sigma, \phi)$.

(iv) For every $(s, \psi) \in \Omega$ there is a neighborhood V of (s, ψ) and a constant M such that $|f_k(\tau, \xi)| \leq M$ for every k and every $(\tau, \xi) \in V$.

(v) The solution of (I) through (σ, ϕ) is unique and is defined in $[\sigma - r, b]$.

Then there exists an index k_0 such that if $k \geq k_0$, every solution x_k of (I_k) through (σ_k, ϕ_k) can be defined in $[\sigma_k - r, b]$ and $x_k \to x$ uniformly in the sense that, given ε, $0 < \varepsilon < b - \sigma + r$, there exists $k_1(\varepsilon)$ such that, if $k \geq k_1$, x_k is defined in $[\sigma - r + \varepsilon, b]$ and $|x_k(x) - x(s)| \leq \varepsilon$ on this interval.

DEFINITION. If x and $\hat{x}$ are solutions os a neutral equation defined in $[\sigma - r, a)$ and $[\sigma - r, b)$, respectively, we say that $\hat{x}$ is a continuation of x if $b > a$ and $x(t) = \hat{x}(t)$ for $t < a$; a solution x *is noncontinuable* if there is no continuation, that is, $[\sigma - r, b)$ is the maximal interval of existence of the solution x.

The continuation theorem has a more general statement than the one we present here, but in application it is used in the following form:

THEOREM. If f takes bounded sets of $\mathbb{R} \times \mathbb{C}$ into bounded sets of $\mathbb{R}^n$, then every bounded noncontinuable solution is defined in a interval of the form $[\sigma, +\infty)$.

If we supress the hypothesis that f takes bounded sets into bounded sets, the last theorem is not true any more, even for retarded differential equations. In fact, if $r > 0$ and $\xi(t)$ is a real bounded function of class C^1 in a bounded interval $[\sigma - r, b)$ such that $\lim_{t \to b} \xi(t)$ does not exist, then, using the Tietze extension theorem, we can find a continuous function $f: \mathbb{C} \to \mathbb{R}$ for which $\xi(t)$ is a solution of $(d/dt) x(t) = f(x_t)$. If, furthermore, $\xi(t)$ is of class C^2, there exists $f: \mathbb{R} \times \mathbb{C} \to \mathbb{R}$, continuous, Lipschitzian in the second variable, such that $\xi(t)$ is solution of $(d/dt) x(t) = f(t, x_t)$ [7].

Given a neutral equation $(d/dt) D(t, x_t) = f(t, x_t)$, if we assume uniqueness of solutions, then the theorems above show

FUNCTIONAL DIFFERENTIAL EQUATIONS

that if $t \geq \sigma$, then the domain of the function T defined by $T\phi = x_t(\sigma, \phi)$ is open and T is continuous in this set.

3. FUNCTIONAL DIFFERENTIAL EQUATION IN $W_{1,\infty}$

The equation

$$\frac{d}{dt}\left[x(t) - x(t-1)\right] = x(t-1)$$

satisfies all the conditions considered in the theory above. Nevertheless, if we write this equation in the form

$$\dot{x}(t) = \dot{x}(t-1) + x(t-1),$$

we see that the space $\mathbb{C}$ cannot be used any more.

We could try the space C^1, but then a simple analysis shows that if the initial data ϕ does not satisfy the condition $\dot{\phi}(0) = \dot{\phi}(-1) + \phi(-1)$, then for $\sigma = 0$ and $t = 1/2$, the function T defined above does not take C^1 into C^1, because $T\phi$ has a discontinuous derivative in $\theta = -1/2$.

A space in which equations of the form

$$\dot{x} = f(t, x_t, \dot{x}_t)$$

can be built in such a way that the operator T defined above preserves the space is $W_{1,\infty}$ [8]. This space can be used to advantage even in the case of retarded equations. For example, the equation $\dot{x} = a\, x(t - r(x))$ with r Lipschitzian, has no uniqueness of solutions in the space C. This is not true in the space $W_{1,\infty}$.

4. AUTONOMOUS LINEAR EQUATIONS

It is well known that the qualitative behavior of the solutions of an ordinary differential equation, in the neighborhood of a critical point, can be described, in general, by the behavior of the linearized equation. To extend these properties to equations of the neutral type, it is necessary to have a good knowledge of the structure of the solution space of the linear equation. Consider, then, a linear autonomous equation of the neutral type:

$$\frac{d}{dt} Dx_t = L(x_t) \quad \text{with} \quad D, L: C = C([-r, 0], E^n) \to E^n, \qquad (1)$$

where E^n is the real or complex n-dimensional space, D is atomic at $\beta = 0$, with a jump at the origin equal to a nonsingular matrix which can be taken equal to the identity. The functional L and D can be represented by the Stieltjes integrals:

$$D\phi = \phi(0) = \int_{-r}^{0} d\mu(\theta) \phi(\theta); \quad L(\phi) = \int_{-r}^{0} d\eta(\theta) \phi(\theta); \quad \phi \in C,$$

where μ and η are matrix of bounded variation which are equal to zero at $\theta = 0$ and are continuous to the left. The atomicity of D means that μ is continuous at $\theta = 0$, that is,

$$\int_{-\varepsilon}^{0} |d\mu(\theta)| \to 0 \quad \text{when} \quad \varepsilon \to 0_+.$$

Let us assume that μ has no singular part, that is,

$$\int_{-r}^{0} d\mu(\theta) \phi(\theta) = \sum_{k=1}^{\infty} A_k \phi(-\omega_k) + \int_{-r}^{0} A(\theta) \phi(\theta) d\theta,$$

where

$$0 < \omega_k < r \quad \text{and} \quad \sum_{k=1}^{\infty} |A_k| + \int_{-r}^{0} |A(\theta)| d\theta < \infty.$$

If $T(t): C \to C$ is defined as $T(t)\phi = x_t(\phi)$, where $x(t, \phi)$ is a solution of (1) satisfying $x_0 = \phi$, then $\{T(t), t \geq 0\}$ is a strongly continuous semigroup of linear operators. The infinitesimal generator A is given by

$$A\phi = \lim_{t \to 0_+} \frac{1}{t} [T(t)\phi - \phi],$$

and the domain $\mathcal{D}(A)$ is the set of $\phi \in C$ for which this limit exists (in the space C).

THEOREM.

(i) $\mathcal{D}(A) = \{\phi \in C \mid \phi' \in C \text{ and } D(\phi') = L(\phi)\}$ and for $\phi \in \mathcal{D}(A)$, $(A\phi)(\theta) = \phi'(\theta)$, $-r \leq \theta \leq 0$.

(ii) $\mathcal{D}(A)$ is dense in C, $T(t)\mathcal{D}(A) \subset \mathcal{D}(A)$, and for ϕ in $\mathcal{D}(A)$, $(d/dt) T(t)\phi = T(t) A\phi = A T(t)\phi$.

(iii) The spectrum of A is a point spectrum and consists of the complex λ which satisfy the characteristic equation

$$\det \Delta(\lambda) = 0,$$

where

$$\Delta(\lambda) = \lambda I - \lambda \int_{-r}^{0} e^{\lambda \theta} d\mu(\theta) - \int_{-r}^{0} e^{\lambda \theta} d\eta(\theta). \qquad (2)$$

The structure of the point and residual spectrum of T(t) is given by (iii) together with a result from the general theory of semigroups. For our particular semigroup T(t) arising from a neutral equation, it can eventually have continuous spectrum (besides zero) but there is no complete characterization for the continuous one, since we do not know if there are points in the continuous spectrum which does not belong to the closure of T(t). The best characterization of the spectrum of a neutral equation is given by Henry [8], which proves the following theorem:

THEOREM. Let α be a real number and suppose that there exists $\varepsilon > 0$ such that in the strip $\alpha - \varepsilon \leq \text{Re } \lambda \leq \alpha + \varepsilon$ there is no characteristic root of (1). Then there is a decomposition $C = P \oplus Q$, with projections E_P and E_Q, P and Q invariants by $T(t)$, and constants M_1, M_2, and $\delta > 0$ such that the restriction $T_P(t)$ of $T(t)$ to P can be extended to a group $\{T_P(t), -\infty < t < \infty\}$ and $|T_Q(t)| \leq M_1 \exp(\alpha - \delta) t$ for $t \geq 0$ and $|T_P(t)| \leq M_2 \times \exp(\alpha + \delta) t$ for $t \leq 0$. In particular, the spectral radius of $T(t)$ is equal to $\exp(at)$, where $a = \sup\{\text{Re } \lambda \mid \det \Delta(\lambda) = 0\}$. Furthermore, if a_D is the spectral radius of $T_D(t)$, corresponding to the equation $(d/dt) D(x_t) = 0$, then $a_D = \sup\{\text{Re } \lambda \mid \det D(e^{\lambda} I) = 0\}$.

This theorem allow us to prove the so-called

POINT SADDLE PROPERTY [5]. Consider the equation

$$\frac{d}{dt} D(x_t) = L(x_t) + F(t, x_t), \qquad (3)$$

which is a perturbation of the linear equation (1) satisfying the condition of that equation with $\alpha = 0$. Suppose that F is a continuous function from $R \times C \to E^n$ satisfying $F(t, 0) = 0$, $|F(t, \phi) - F(t, \psi)| \leq \mu(\sigma) |\phi - \psi|$, for $|\phi|, |\psi| \leq \sigma$, where $\mu(\sigma) \to 0$ when $\sigma \to 0_+$. Then there exists $\sigma > 0$, such that:

(a) If $S = \{\phi \in C \mid |E_Q Q| < \sigma/2M_1$ and $|x_t(s, \phi)| \leq \sigma$ for every $t \geq s\}$, then E_Q restricted to S is a homeomorphism from S to $\{\phi \in Q \mid |\phi| \leq \sigma/2M_1\}$. We have also that S is tangent to Q at the origin, and every solution with initial value in S goes to zero exponentially when $t \to +\infty$.

(b) If $U = \{\phi \in C \mid |E_P \phi| \leq \sigma/2M_2$ and $|x_t(s, \phi)| \leq \sigma$ for every $t \leq s\}$, then E_P restricted to U is a homeomorphism from U into $\{\phi \in P \mid |\phi| \leq \sigma/2M_2\}$. Furthermore, U is tangent to P at the origin and every solution with initial value in U goes to zero exponentially when $t \to -\infty$.

FUNCTIONAL DIFFERENTIAL EQUATIONS

Remark. The above theorem is also true for the more general perturbed equation of the type

$$\frac{d}{dt}\left[D(x_t) - G(t, x_t)\right] = L(x_t) + F(t, x_t), \qquad (4)$$

which is not stated here since we are dealing with the case in which D is linear.

5. ASYMPTOTIC BEHAVIOR

Because of technical difficulties and the necessity of a deeper knowledge of the structure of linear autonomous systems, which is made difficult because of the existence of the continuous spectrum, there are very few results about the asymptotic behavior of functional differential equations of the neutral type. We can be state, in particular, the following theorem [9], which generalizes results by Hale for retarded equations.

Let μ, Re $\mu > a_D$ be a simple characteristic root of (2), that is, μ is a simple root of det $\Delta(\lambda) = 0$. Then there exists a column vector c and a row vector d such that $ce^{\mu t}$ is a solution of (2) and $de^{-\mu \tau}$ is a solution of the adjoint equation of (2):

$$\frac{d}{d\sigma}\left[y(\sigma) - \int_{-r}^{0} y(\sigma - \theta) \, d\mu(\theta)\right] = -\int_{-r}^{0} y(t - \theta) \, d\eta(\theta).$$

Let

$$\phi_\mu(\theta) = ce^{\mu\theta}, \quad -r \leq \theta \leq 0; \quad \psi_\mu(\tau) = de^{-\mu\tau}, \quad 0 \leq \tau \leq r;$$

$$\alpha^{-1} = (\psi_\mu, \phi_\mu)$$

$$= \psi_\mu(0) \, D(\phi_\mu) + \int_{-r}^{0} \int_{0}^{\theta} \psi_\mu(\xi - \theta) \left[d\mu(\theta)\right] \phi_\mu(\xi) \, d\xi$$

$$- \int_{-r}^{0} \int_{0}^{\theta} \psi_\mu(\xi - \theta) \left[d\eta(\theta)\right] \phi_\mu(\xi) \, d\xi. \qquad (5)$$

Let $X_0(\theta) = 0$, $-r \leq \theta \leq 0$, $X_0(\theta) = 1$, $\theta = 0$.

THEOREM. Suppose that Re $\mu > a_D$, where μ is a simple characteristic root of

$$\det \Delta(\lambda) = 0, \qquad (6)$$

$$\Delta(\lambda) = I - \int_{-r}^{0} e^{\lambda\theta} d\mu(\theta) - \int_{-r}^{0} e^{\lambda\theta} d\eta(\theta)$$

$$= \lambda D(e^{\lambda \cdot} I) - L(e^{\lambda \cdot} I),$$

and assume that all the other characteristic roots of (6), with real part equal to Re μ, are also simple. Let ϕ_μ and α be defined as in (5). Let

$$\delta(t) = \alpha \, d\{F(t, \phi_\mu) + \mu \, G(t, \phi_\mu) + F[t, X_0 \, G(t, \phi_\mu)]\},$$

where $F(t, \phi)$ and $G(t, \phi)$ are linear in ϕ, and assume that there exist continuous functions $\gamma(t)$ and $\pi(t)$ for which $F(t, \phi)$ and $G(t, \phi)$ satisfy

$$|F(t, \phi)| \leq \gamma(t) \|\phi\|_1, \quad |G(t, \phi)| \leq \pi(t) \|\phi\|_1, \quad t \geq \sigma.$$

Let $s(t, \sigma) = \int_\sigma^t \delta(\xi) \, d\xi$, and let $\gamma(t)$ and $\pi(t)$ satisfy one of the following hypotheses:

(I) $\int^\infty \gamma(t) \, dt < \infty$, $\int^\infty \pi(t) \, dt < \infty$, $\lim_{t\to\infty} \pi(t) = 0$.

(II) For some $\beta > 0$ there exists a function $\gamma_2(t)$ defined for $t \geq 0$ such that

$$\lim_{t\to\infty} \pi(t) = 0, \quad \lim_{t\to\infty} \gamma_2(t) = 0;$$

$$\int^\infty \gamma(t) \gamma_2(t) \, dt < \infty, \quad \int^\infty \pi(t) \gamma_2(t) \, dt < \infty;$$

FUNCTIONAL DIFFERENTIAL EQUATIONS

$$\int_\sigma^t \exp[-\beta(t - \tau) - \operatorname{Re} s(t, \tau)]$$

$$\times [\gamma(\tau) + \pi(\tau) + \gamma(\tau)\pi(\tau)]\, d\tau \leq \gamma_2(t), \quad t > \sigma \geq 0;$$

$$\int_t^\infty \exp[\beta(t - \tau) - \operatorname{Re} s(t, \tau)]$$

$$\times [\gamma(\tau) + \pi(\tau) + \gamma(\tau)\pi(\tau)]\, d\tau \leq \gamma_2(t), \quad t \geq 0;$$

$$\int_{t+\theta_2}^{t+\theta_1} \exp[-\beta(t - \tau) - \operatorname{Re} s(t, \tau)]$$

$$\times [\gamma(\tau) + \pi(\tau) + \gamma(\tau)\pi(\tau)]\, d\tau \leq \gamma_2(t)|\theta_1 - \theta_2|, \quad -r \leq \theta_1, \theta_2 \leq 0.$$

Assume also that for any real number ν, for any n-dimensional row vectors η and λ, and for any continuous function W_t defined for $t \geq \sigma$ with values in C' for which $\|W_t\|$ is bounded when $t \to \infty$, we have

$$\left| \int_\sigma^\infty \exp[i\nu(t - \tau) - s(t, \tau)] \right.$$

$$\left. \times \{\eta[F(\tau, W_\tau) + F(\tau, X_0 G(\tau, W_\tau))] - \lambda G(\tau, W_\tau)\}\, d\tau \right|$$

$$\leq [|\eta| + |\lambda|]\, \gamma_2(t) \sup_{\tau \geq \sigma} \|W_\tau\|_1, \quad t \geq \sigma.$$

Then there exists $\sigma > 0$ sufficiently large and a vector $a \neq 0$ such that the system

$$\frac{d}{dt}[D(x_t) - G(t, x_t)] = L(x_t) + F(t, x_t) \qquad (4)$$

has a solution $x(t)$ defined for $t \geq 0$ satisfying

$$x(t) = \exp[\mu(t - \sigma) + s(t, \sigma)][a + o(1)]$$

when $t \to \infty$.

6. STABILITY

DEFINITION. The solution $x = 0$ of the neutral equation

$$\frac{d}{dt} D(t) x_t = f(t, x_t) \tag{7}$$

is uniformly asymptotically stable in $[0, +\infty)$ if, given $\varepsilon > 0$, there exists $\delta(\varepsilon) > 0$ such that $|\phi| \le \delta(\varepsilon)$ implies $|x_t(\sigma, \phi)| \le \varepsilon$ for $t \ge \sigma \ge 0$ and there exists $\eta > 0$ such that, given $\varepsilon > 0$, there exists $T(\varepsilon)$ satisfying $|x_t| \le \varepsilon$ if $t \ge \sigma + T(\varepsilon)$ and $|\phi| \le \eta$.

If the last condition is true for every η (in this case T depends also on η), we say that the solution $x = 0$ is globally uniformly asymptotically stable.

A basic tool in the investigation of the stability of equations obtained by perturbations of a linear nonautonomous equation is the formula of variation of constants. Consider the equation

$$\frac{d}{dt} D(t, x_t) = L(t, x_t) + F(t, x_t), \tag{8}$$

considered as a perturbation of the linear equation

$$\frac{d}{dt} D(t, x_t) = L(t, x_t). \tag{9}$$

If the solution $x_t(\sigma, \phi)$ of (9) is denoted by $T(t, \sigma) \phi$, then there exists a matrix of functions $n \times n$, $B(t, s)$ defined for $0 \le s \le t + r$, $t \in [0, \infty)$, continuous to the right with respect to s, of bounded variation in s, $B(t, s) = 0$, $t \le s \le t + r$, such that the solution $x_t(\sigma, \phi)$ of (8) satisfies

$$x_t(\sigma, \phi) = T(t, \sigma) \phi + \int_\sigma^t B_t(\cdot, s) F(s, x_s) ds, \quad t \ge \sigma;$$

furthermore, if the solution of (9) is uniformly asymptotically

FUNCTIONAL DIFFERENTIAL EQUATIONS

stable, there exist constants $M \geq 1$ and $\alpha \geq 0$ for which

$$|T(t, \sigma)| \leq Me^{-\alpha(t-\sigma)}, \quad t \geq \sigma;$$

$$|B_t(\cdot, s)| \leq Me^{-\alpha(t-s)}, \quad t \geq s;$$

$$\int_\sigma^s |du\, B_t(\cdot, u)| \leq Me^{-\alpha(t-s)}, \quad t \geq s \geq \sigma.$$

With the above results, almost all theorems on perturbations of ordinary differential equations can be extended to neutral equations. For example: [10], if $F = F_1 + F_2$ and F_1 and F_2 are "small" (the way in which F_1 is small may not be the same for F_2), then the solution $x = 0$ of the perturbed equation is uniformly asymptotically stable; this theorem is true even we perturb also the operator D (see [10] for more details).

7. LYAPUNOV THEORY

To study the stability of a neutral equation using Lyapunov functionals, we have to restrict ourselves to a particular class of operators $D(t)$.

DEFINITION. We say that $D(t)$ is uniformly stable in $[0, \infty)$ if the function $|D(t)|$ is bounded in this interval and there exist constants K and $\alpha > 0$ such that the solution of $D(t, x_t) = 0$, $x_\tau = \phi$, $D(\sigma, \phi) = 0$ satisfies $|x_t| \leq Ke^{-\alpha(t-\sigma)}|\phi|$ for $t \geq \sigma \geq 0$.

For example, if $D\phi = \phi(0) - \Sigma_{i=1}^{N} A_i \phi(-\omega_i)$, $0 < \omega_i \leq r$, where A_i is a matrix n x n, then D is uniformly stable if there exist $\varepsilon > 0$ such that the solutions of

$$\det \Delta(\lambda) = \det\left(I - \sum_{i=1}^{N} A_i e^{-\omega_i \lambda}\right) = 0$$

satisfy $\operatorname{Re} \lambda \leq -\varepsilon$.

DEFINITION. If $V: [0, \infty) \times C \to R$ is a continuous function, then the derivative $\dot{V}(\sigma, \phi)$ of V at the point (σ, ϕ) along the solutions of the neutral equation $(d/dt) D(t) x_t = f(t, x_t)$ is

$$\dot{V}(\sigma, \phi) = \limsup_{t \to \sigma_+} \frac{V(t, x_t) - V(\sigma, \phi)}{t - \sigma}.$$

THEOREM [11]. Suppose that $D(t)$ is uniformly stable and there exists a neighborhood U of the origin in the space C such that $f([0, \infty) \times U)$ is a bounded set. Suppose also that there is a continuous function $V: [0, \infty) \times C \to R$ and functions $u(s)$ and $v(s)$, $0 < s < H$, such that:

(i) $u(|D(t)\phi|) \leq V(t, \phi) < v(|\phi|)$.

(ii) u and v are continuous and nondecreasing; $u(s) < v(s)$ and $u(s) > 0$ if $s > 0$.

(iii) $\dot{V}(\sigma, \phi) \leq -\omega(|D(t)| \phi)$, where $\omega(s)$ is continuous function in $[0, H]$, with $\omega(s) > 0$ if $s > 0$.

Then the solution $x = 0$ is uniformly asymptotically stable. Furthermore, if $H = +\infty$ and $u(s) \to \infty$ when $s \to \infty$, then $x = 0$ is globally uniformly asymptotically stable.

In applications we use functionals of the following kind:

$$V(t, \phi) = \beta(D(t) \phi) + \int_{-r}^{0} \alpha(\phi(\theta)) \, d\theta,$$

with $\alpha, \beta: \mathbb{R}^n \to R$ continuous and β of class C^1, since for this case is easy to see that

$$\dot{V}(t, \phi) = \text{grad } \beta(D(t) \phi) \times f(t, \phi) + \alpha(\phi(0)) - \alpha(\phi(-r)).$$

As an example, consider the following equation [12]:

$$C \frac{d}{dt} \left[x(t) + K x(t - r) \right]$$

$$= -\frac{x(t)}{Z} + \frac{K}{Z} x(t - r) - g(x(t)) - K g(x(t - r)),$$

$$C > 0, \quad Z > 0, \quad |K| < 1.$$

If we define

$$V(\phi) = \frac{1}{2} (D\phi)^2 + \alpha \int_{-r}^{0} \phi^2(\theta) \, d\theta, \quad \alpha > 0,$$

$$u(s) = \frac{s^2}{2}, \quad \text{and} \quad v(s) = \left(\frac{(1 + |K|)^2}{2} + \alpha r \right) s^2,$$

we see that conditions (i) and (ii) of the theorem are satisfied.

We have also that

$$\dot{V}(\phi) = -\left(\frac{1}{CZ} + \frac{g(\phi(0))}{C\phi(0)} - \alpha \right) \phi^2(0)$$

$$- \left(\frac{K}{C} \frac{g(\phi(-r))}{\phi(-r)} - \frac{K}{CZ} + \frac{K}{CZ} + \frac{K}{C} \frac{g(\phi(0))}{\phi(0)} \right) \phi(0) \phi(-r)$$

$$- \left(-\frac{K^2}{CZ} + \frac{K^2}{C} \frac{g(\phi(-r))}{\phi(-r)} \right) \phi^2(-r).$$

If we can write $\dot{V}(\phi)$ in the form

$$\dot{V}(\phi) = -A(D\phi)^2 - B\phi^2(0) - 2G \phi(0) \phi(-r) - J \phi^2(-r),$$

where A is constant, $A > 0$, and B, G, J are functions such that $BJ \geq G^2$, then $\dot{V}(\phi) \leq -A(D\phi)^2$, and condition (iii) of the theorem is satisfied and the solution $x = 0$ will be globally uniformly asymptotically stable. A sufficient condition to have V satisfying the above condition is given as follows concerning g:

$$\left(\frac{g(\rho)}{\rho}\right)_M < \left(\frac{1 - |K|}{1 + |K|}\right) \frac{1}{Z} + \left(\frac{g(\rho)}{\rho}\right)_m$$

and

$$\left(\frac{g(\rho)}{\rho}\right)_m \geq -\frac{1}{Z} \left(\frac{1 - |K|}{1 + |K|}\right),$$

where

$$\left(\frac{g(\rho)}{\rho}\right)_M = \sup_{\rho \geq 0} \frac{g(\rho)}{\rho} \quad \text{and} \quad \left(\frac{g(\rho)}{\rho}\right)_m = \inf_{\rho \geq 0} \frac{g(\rho)}{\rho}.$$

8. EXISTENCE OF FORCED OSCILLATIONS

If a neutral equation (7) has ω-periodic coefficients, that is, if D and f are ω-periodic in the variable t, then, assuming uniqueness of solutions, it is easy to see that the existence of an ω-periodic solution is equivalent to the existence of a fixed point of the transformation $T_\omega(\phi) = x_\omega(0, \phi)$.

DEFINITION. The neutral equation (7) is uniformly eventually bounded if there exist a constant R_0 and functions $\alpha(R)$ and $T(R)$ such that $|x_t(\sigma, \phi)| \leq \alpha(R)$ if $|\phi| \leq R$ and $t \geq \sigma$, and $|x_t(\sigma, \phi)| \leq R_0$ if $t \geq \sigma + T(R)$.

A generalization of an asymptotic fixed-point theorem due to Browder, for a more general class of functions than the completely continuous or compact one, allows us to prove the following theorem:

THEOREM [13]. If the neutral equation (7) has periodic coefficients and is uniformly eventually bounded, $D(t)$ is uniformly stable, and f takes bounded sets of $R \times C$ into bounded sets of $\mathbb{R}^n$, then (7) has ω-periodic solutions.

The verification of the eventual uniform boundedness is done using techniques that generalize the Razumikhin theorem for retarded equations. As an example of application, we mention that the neutral equation

$$\frac{d}{dt}\left[x(t) - g\,x(t-s)\right]$$

$$= a\,x(t) + b\,x(t-1) - \gamma\,x^3(t) + g\gamma\,x^3(t-1) + p(t)$$

with $|g| < 1$, $\gamma > 0$ and $p(t)$ continuous and ω-periodic, has periodic solutions [13].

REFERENCES

1. J. A. Yorke, Selected topics in differential delay equations, Japan-U.S.A. Seminar on Ordinary Differential and Functional Equations; *Lecture Notes in Mathematics*, No. 243, Springer Verlag, New York, 1971, pp. 16-28.
2. R. K. Brayton, Nonlinear oscillations in a distributed network, *Quart. Applic. Math.*, 24 (1967), 289-301.
3. M. Slemrod, Non existence of oscillations in a nonlinear distributed network, *J. Math. Anal. Appl.*, 36 (1971), 22-40.
4. M. A. Krasnoselskii, *Topological Methods in the Theory of Nonlinear Integral Equations*, Macmillan, New York, 1964.
5. M. A. Cruz and J. K. Hale, Existence, uniqueness, and continuous dependence for hereditary systems, *Ann. Mat. Pura Appl.* 85 (1970), 63-82.
6. J. K. Hale, Forward and backward continuation of NFDE's, *J. Diff. Eq.*, 9 (1971), 168-181.
7. J. A. Yorke, *Proc. Amer. Math. Soc.*, 21, No. 2 (1969), 648-652.

8. D. Henry, Linear autonomous neutral functional differential equations, *J. Diff. Eq.*, *15* (1976), 106-128.

9. A. F. Izé and N. A. Molfetta, Asymptotically autonomous neutral functional differential equations with time dependent lag, *J. Math. Anal. Appl.*, *51*, No. 2 (1975), 299-325.

10. J. K. Hale and A. F. Izé, On the uniform asymptotic stability of functional differential equations of the neutral type, *Proc. Amer. Math. Soc.*, *28*, No. 1, (1971), 100-106.

11. M. A. Cruz and J. K. Hale, Asymptotic behavior of NFDE, *Arch. Rat. Mech. Anal.*, *34* (1969), 331-353.

12. O. F. Lopes, Forced oscillations in NFDE, (submitted to *SIAM J. Appl. Math.*).

13. J. K. Hale and O. F. Lopes, Dissipative processes and asymptotic fixed-point theorems, *J. Diff. Eq.*, March (1973), 391-402.

ON THE GEOMETRY AND STRUCTURE OF BANACH SPACES[*]

L. A. Karlovitz

Institute for Fluid Dynamics and Applied Mathematics
University of Maryland
College Park, Maryland

1. INTRODUCTION

We are concerned with some geometric aspects of Banach spaces at two opposite ends of the spectrum. On the one hand we investigate the geometry and structure of a class of nonreflexive Banach spaces which have only recently been defined. On the other hand, we study some classical variational problems in Hilbert space, which are seen to have a considerable amount of geometric content.

In Section 2 we discuss *flat* Banach spaces. This is a class of nonreflexive spaces which are characterized by strong, albeit unusual, geometric properties, some of which justify the term *flat*. We give a survey of results about flat spaces. They include results about their geometry, tree properties (in the sense of R. C. James), subspace structure, and Schauder bases.

---

[*] This work was done in part while the author was visiting at the University of Brasília.

In Section 3 we discuss four classical variational problems which are set in Hilbert space. Two are variational principles for eigenvalues, and two are Kolmogorov n-width problems. We discuss the interrelation of the problems and emphasize their geometric and approximation theoretic aspects. We point out certain confusions which have existed in the past.

Most of the results are of recent vintage and easily accessible. Hence proofs are not given here.

2. FLAT BANACH SPACES

A Banach space X, with norm denoted by $\|\cdot\|$, is said to be *flat* if there exists a function $g: [0, 2] \to X$ such that

$$g(s) = 1 \text{ for each } s \in [0, 2], \quad g(0) = -g(2), \tag{1}$$

and g is Lipschitz continuous with constant 1.

This is equivalent to the statement that the *girth* of the unit ball of X (defined by Schäffer [1] to be the infimum of the lengths of all rectifiable centrally symmetric curves which lie in the surface of the unit ball) is 4 *and* is achieved by some curve (i.e., the infimum is achieved as a minimum). The function g of (1) is easily seen to be the arc-length representation of a curve lying in the surface of the unit ball with antipodal endpoints and with length 2. This curve together with its reflection through the origin achieves the girth 4. These spaces were introduced and initially analyzed by Harrell-Karlovitz [2-4]. They have distinguished geometric properties, as a consequence of which the term *flat* if introduced in [3]. It follows from any one of several of the properties that flat Banach spaces are necessarily nonreflexive. In particular, it

ON THE GEOMETRY AND STRUCTURE OF BANACH SPACES

follows from the lack of weak differentiability of the function g (Theorem 2 below).

A Banach space X is said to be *completely flat* if there exists a function g: $[0, 2] \to X$ satisfying (1) and

$$X = \text{closed linear hull } \{g(s): s \in [0, 2]\}. \quad (2)$$

Flat Banach spaces and spaces isomorphic thereto comprise a large proper subset of all nonreflexive spaces. For example, $L_1[0, 1]$ is completely flat, and $L_1[\mu]$ is flat if and only if μ is not purely atomic. $C[0, 1]$ is flat and contains a completely flat subspace of co-dimension 1, namely all functions satisfying $f(0) = -f(1)$. More generally, $C_\sigma(K)$ is flat for various choices of σ and K. The new space of Lindenstrauss-Stegall [5] is isomorphic to a completely flat space. These examples are treated in Harrell-Karlovitz [2, 3, 6], Nyikos-Schäffer [7], and Schäffer [8]. The dual of a flat Banach space is again flat (Theorem 4 below). An interesting comparison is the following. A space isomorphic to a flat Banach space fails to have the Radon-Nikodym property (for example, Huff [9]), but not conversely. We give some details for $L_1[0, 1]$ in Examples 1 and 3 below, and for $C[0, 1]$ in Example 2.

Example 1. For an arbitrary point $y \in L^1[0, 1]$ such that $\|y\| = 1$, define $g_y: [0, 2] \to L^1[0, 1]$ by

$$g_y(s) = \begin{cases} -y(t), & t \in [0, \alpha] \\ y(t), & t \in (\alpha, 1] \end{cases}, \text{ where } \int_0^\alpha |y(t)| \, dt = \frac{s}{2}$$

It is easy to see that g_y satisfies (1). Moreover if $y(t) \equiv 1$, then $L^1[0, 1] = $ closed linear hull $\{g_y(s): s \in [0, 2]\}$. In general, it can be shown that the closed linear hull $\{g_y(s): s \in [0, 2]\}$ is isometrically isomorphic to $L^1[0, 1]$.

If X is a flat Banach space and g: $[0, 2] \to X$ satisfies (1), we say that g is a *girth curve* for X. If g is a girth curve for X, then for each pair of reals s, h with $s \in (0, 2]$ and $h \in (0, s]$ we define the *difference quotient* $\Delta_g(s, h)$ by

$$\Delta_g(s, h) = h^{-1}(g(s - h) - g(s)). \qquad (3)$$

Further, for each $s \in [0, 2]$ we define the *chord set of g at s*, denoted by $\chi(g, s)$, by

$$\chi(g, s) = \text{closed convex hull } (\{-\Delta_g(t, h): t \in (0, s], h \in (0, t]\}$$

$$\bigcup \{\Delta_g(t, h): t \in (s, 2], h \in (0, t - s]\}). \qquad (4)$$

If $s = 0$ ($s = 2$), the first (second) set is empty. Finally, for each $s \in [0, 2]$ we choose a linear functional $f_g^*(s) \in X^*$, the dual space of X, which satisfies

$$\|f_g^*(s)\| = 1, \quad <f_g^*(s), g(s)> = 1, \qquad (5)$$

where $<\cdot, \cdot>$ denotes the pairing between X and X^*. The following theorem and remark describe the special properties of $\chi(g, s)$ and justify the terms *flat Banach space* and *completely flat Banach space*.

THEOREM 1. ([4, Theorem 1]) Let X be a flat Banach space. Let g: $[0, 2] \to X$ be a girth curve for X, i.e., g satisfies (1). Then for each $s \in [0, 2]$

$$g(s) \in \chi(g, s) \subset \{x: <f_g^*(s), x> = 1\} \cap \{x: \|x\| = 1\}. \qquad (6)$$

Moreover, for each $y \in \chi(g, s)$,

$$\sup_{x \in \chi(g,s)} \|y - x\| = \sup_{x, z \in \chi(g,s)} \|x - z\| = 2, \qquad (7)$$

i.e., the diameter of $\chi(g, s)$ is 2 and each point of $\chi(g, s)$ is diametral. (Hence $\chi(g, s)$ fails to have normal structure.) Finally, if X = closed linear hull $\{g(t): t \in [0, 2]\}$, then

$$\text{closed affine hull } \chi(g, s) = \{x: \langle f_g^*(s), x \rangle = 1\}. \qquad (8)$$

Remark 1. According to (6), $\chi(g, s)$ is a *flat* area in the surface of the unit ball, i.e., it belongs to the intersection of the unit ball with a supporting hyperplane. Since each subset of the unit ball has diameter at most 2, $\chi(g, s)$ is, by (7), maximal in the sense of diameter. (Moreover, each of its points is diametral.) Thus if we imagine a circumnavigation of the unit ball along the path formed by $g: [0, 2] \to X$ and $-g: [0, 2] \to X$, then we observe two properties: the length of the journey is only 4, and each point of it lies in a large flat area. Hence the term *flat Banach space*. If, moreover, X = closed linear hull $\{g(s): s \in [0, 2]\}$ then, by (8), the flat area looks "locally" like a hyperplane. Hence the term *completely flat Banach space* (or the alternate *pre-Columbian Banach space* suggested by R. J. Duffin).

Remark 2. According to Example 1, each point $y \in L_1[0, 1]$ such that $\|y\| = 1$ is the initial point of some girth curve, i.e., $g_y(0) = y$. Therefore each point y in the surface of the unit ball not only fails to be an extreme point, but it belongs to a large flat area $\chi(g_y, 0)$ which satisfies the conclusions on Theorem 1.

As can be expected, a path of circumnavigation of the unit ball having length 4 is highly nondifferentiable. This is expressed by the following theorem. The second part of the theorem relies on a result of Gelfand [10, Theorem 3] concerning the differentiability of Lipschitz continuous functions with values in a separable dual space.

THEOREM 2. ([4, Theorem 4]) Let X be a flat Banach space. Let g: $[0, 2] \to X$ be a girth curve for X. Then g fails to be weakly differentiable for each $s \in [0, 2]$. It follows that X is not isomorphic to any subspace of any separable dual space.

Thus, in particular, flat spaces are nonreflexive; but there exist nonreflexive spaces, for example ℓ_1, which are not isomorphic to any flat space.

Girth curves have an interesting *replication* property. Namely, if g satisfies (1) and $\Delta_g(s, h)$ is given by (3) then for each pair of positive integers j and n, with $1 \leq j \leq 2^n$ we readily observe

$$g(j/2^n) = (-1/2^{n+1}) \sum_{i=1}^{j} \Delta_g(i/2^n, 1/2^n)$$
$$+ (1/2^{n+1}) \sum_{i=j+1}^{2^{n+1}} \Delta_g(i/2^n, 1/2^n).$$

Thus $g(j/2^n)$ is a combination of elements of $X(g, 0)$. In other words, the entire curve g can be replicated by simple formulas from elements of $X(g, 0)$, which lies entirely in one of the supporting planes of the curve. Similarly, it can be replicated from $X(g, s)$ for each $s \in [0, 2]$.

This fact can be expressed in somewhat different terms. To this end let $x((2i - 1)/2^n) = \Delta_g(i/2^{n-1}, 1/2^{n-1})$, $n = 1, 2, \ldots$, $i = 1, \ldots, 2^{n-1}$. Then by (2),

$$x((2i - 1)/2^n) = (1/2) \, x((4i - 3)/2^{n+1})$$
$$+ (1/2) \, x((4i - 1)/2^{n+1}). \qquad (9)$$

Furthermore, by (6), for each $s \in [0, 2]$

$$\langle f_g^*(s), x((2i - 1)/2^n) \rangle = \begin{cases} -1, & i/2^{n-1} \leq s \\ \\ 1, & s \leq (i - 1)/2^{n-1} \end{cases} \qquad (10)$$

It follows, in particular, that X has the *infinite tree property* of James [11] because the set of points $x((2i - 1)/2^n)$ form an infinite tree. By (6) all of the points lie in the unit ball; they "branch" according to (9); and, as a consequence of (10), $\|x((4i - 3)/2^{n+1}) - x((4i - 1)/2^{n+1})\| \geq 2$. The space X also satisfies the *infinite supported tree property* of Harrel-Karlovitz [6] because (10) is satisfied; i.e., for each rational $k/2^m \in [0, 1]$, the infinite tree is *supported* by the hyperplanes $\{x: <f_g^*(k/2^m), x> = 1\}$ and $\{x: <f_g^*(k/2^m), x> = -1\}$; i.e., each point of the tree lies in one of the two hyperplanes or it is a finite convex combination of ones that do. This property gives an isomorphic characterization of flat spaces.

THEOREM 3. ([6, Theorem 1]) The Banach space X is isomorphic to a flat Banach space Y if and only if X has the infinite supported tree property. If X has this property then its dual space X^* also has it.

The second part of the theorem shows that the infinite supported tree property is strictly stronger than the infinite tree property. For the latter is not preserved under duality. For example, c_0 has the infinite tree property, while ℓ_1 does not (James [11]).

The second part of the theorem is a corollary of the first part and of the following theorem.

THEOREM 4. ([12, Theorem 3]) If X is a flat Banach space then its dual space X^* is also flat. Moreover, if X is completely flat, then its dual X^* has a girth curve $g^*: [0, 2] \to X^*$ with the additional property: for each $s \in [0, 2]$, the functional $g^*(s)$ fails to achieve its supremum over the unit ball of X, i.e., $\|g^*(s)\| = 1$ but $|<g^*(s), x>| < 1$ for each $x \in X$, $\|x\| \leq 1$.

The function g^* is explicitly constructed in [12]. Hence, for this class of spaces, this provides a new construction for functionals which fail to achieve their suprema. Furthermore, there is a multiplicity of such functions g^*. Thus the construction yields new information about the location of such functionals.

We now turn to the subspace structure of flat Banach spaces.

THEOREM 5. ([13, Theorem 1]) Let X be flat Banach space. Let $g: [0, 2] \to X$ be a girth curve for X. Suppose that the functional

$$|\cdot| = \sup_{s \in [0,2]} |<f_g^*(s), \cdot>|,$$

defines an equivalent norm on X. Then X contains a subspace isomorphic to ℓ_1 and a subspace isomorphic to c_0.

Completely flat Banach spaces are clearly separable, and the dual of any flat space is nonseparable (Karlovitz [12]). Hence one might expect every flat Banach space to contain a subspace isomorphic to ℓ_1. However, the aforementioned new space of Lindenstrauss-Stegall [5] is separable, has nonseparable dual and fails to contain a subspace isomorphic to ℓ_1; moreover, it is isomorphic to a completely flat Banach space (Harrell-Karlovitz [6]).

It is conjectured that each flat Banach space *either* contains a subspace isomorphic to ℓ_1 *or* c_0 is finitely representable in it. This is known to be true for a large class of flat spaces (Karlovitz [13]), and the general case is still under investigation.

Some aspects of the subspace structure can also be expressed in terms of trees. This is done in the following two theorems which make use of Hagler [14] and Lindenstrauss-Stegall [5] as well as the above theory of flat spaces.

ON THE GEOMETRY AND STRUCTURE OF BANACH SPACES

THEOREM 6. ([6, Theorem 3]) Suppose that X^* is the dual of a Banach space X. If X contains a subspace isomorphic to ℓ_1 then X^* has the infinite supported tree property, but not conversely.

A somewhat stronger tree property can be used to characterize the duals of spaces containing subspaces isomorphic to ℓ_1. We shall say that a Banach space X has the ℓ_1-*tree property* if there exists a sequence $\{x((2i - 1)/2^n): n = 1, 2, \ldots, i = 1, \ldots, 2^{n+1}\}$ in the unit ball of X so that (9) is satisfied and so that

$$m \sum_{i=1}^{2^{n+1}} |\lambda_i| \le \left\| \sum_{i=1}^{2^{n+1}} \lambda_i x((2i - 1)/2^n) \right\| \le M \sum_{i=1}^{2^{n+1}} |\lambda_i|$$

for some m, M > 0 and all choices of n and λ_i.

THEOREM 7. ([13, Theorem 4]) Suppose that X^* is the dual of a Banach space X. Then X contains a subspace isomorphic to ℓ_1 if and only if X^* has the ℓ_1-tree property.

We now consider the basis problem for flat Banach spaces. For this we note that many flat spaces, including all of the examples above, satisfy a further property; namely, the norm is *monotone* with respect to a girth curve g in the following sense:

$$\left\| \sum_{i=1, i \ne j, j+1}^{n} \lambda_i \Delta_g(s_i, s_i - s_{i-1}) + (\lambda_j + \lambda_{j+1}) \Delta_g(s_{j+1}, s_{j+1} - s_{j-1}) \right\|$$

$$\le \left\| \sum_{i=1}^{n} \lambda_i \Delta_g(s_i, s_i - s_{i-1}) \right\|, \qquad (11)$$

for all real λ_i and all $0 \le s_0 < \cdots < s_n \le 2$. This property is discussed for $C[0, 1]$ and $L_1[0, 1]$ in Examples 2, and 3 below. Our main result on Schauder bases is the following.

THEOREM 8. ([15, Theorem 1]) Let X be a completely flat Banach space such that there exists a girth curve $g: [0, 2] \to X$ satisfying (2) and (11). Then X has a Schauder basis $\{e_k : k = 0, 1, \ldots\}$ which can be constructed as follows: $e_0 = g(0)$ and

$$e_k = 2^{m-1}\left[-g\left(\frac{2(k - 2^m)}{2^m}\right) + 2g\left(\frac{2(k - 2^m) + 1}{2^m}\right) - g\left(\frac{2(k - 2^m) + 2}{2^m}\right)\right]$$

$$= \frac{1}{2}\left[\Delta_g\left(\frac{2(k - 2^m) + 2}{2^m}, \frac{1}{2^m}\right) - \Delta_g\left(\frac{2(k - 2^m) + 1}{2^m}, \frac{1}{2^m}\right)\right],$$

for $k \geq 1$, where the integer m satisfies $2^m \leq k \leq 2^{m+1} - 1$.

This theorem gives a common construction for a Schauder basis for each of a quite diverse class of nonreflexive spaces. For the space $C[0, 1]$ the construction yields the original basis of polygonal functions of Schauder [16] (Example 2 below). Thus it can be viewed as a generalization of his construction. We give two examples.

Example 2. Define $g: [0, 2] \to C[0, 1]$ by

$$g(s)(t) = \begin{cases} 1 - s + 2t, & 0 \leq t \leq s/2, \\ 1 + s - 2t, & s/2 < t \leq 1, \end{cases}$$

It is easily verified that g satisfies (1) and that $C[0, 1]$ = closed linear hull $\{\{1 - g(0)\} \cup \{g(s): s \in [0, 2]\}\}$. Moreover, the two functions appearing in (11) are polygonal and they agree on the points $\{s_i: i \neq j\}$, which are the nodes of the function appearing in the left-hand side. Hence the inequality (11) follows immediately. If the basis is constructed according to Theorem 8, then $\{\{1 - g(0)\} \cup \{e_k: k = 0, 1, \ldots\}\}$ is precisely the polygonal basis of Schauder [16].

Example 3. We apply the formulas of Example 1 to the functions $y_1(t) \equiv 1$, and $y_2(t) = 2t$ to find

$$g_1(s)(t) = \begin{cases} -1, & 0 \le t \le s/2, \\ 1, & s/2 < t \le 1, \end{cases} \qquad g_2(s)(t) = \begin{cases} -2t, & 0 \le t \le \sqrt{s/2} \\ 2t, & \sqrt{s/2} < t \le 1. \end{cases}$$

It is readily seen that g_1 and g_2 satisfy (1) and (2). Moreover, $[g_1(t_2) - g_1(t_1)]/(t_2 - t_1)$ is the normalized characteristic function of the interval $[t_1/2, t_2/2]$, whence (11) follows immediately. For g_2 the verification of (11) is analogous. The construction of Theorem 8 applied to g_1 yields the classical Haar system as a basis. On the other hand, g_2 yields a somewhat less classical basis which is *not* orthogonal. Other bases can be constructed in a similar manner.

Finally we turn to an interesting open problem; namely, to characterize nonreflexivity in terms of the girth of the unit ball. As we have noted above, if X is flat, i.e., if the girth is 4 *and* if it is achieved by some curve, then X is nonreflexive but not conversely, not even isomorphically. This complements a result of Schäffer-Sundaresan [17] which states that if X is nonreflexive then the girth is 4, but not conversely. Thus the unit ball having girth 4 achieved by some curve is too strong to characterize nonreflexivity and just having girth 4 is too weak. The desired intermediate property should be expressible in the following terms. There exists a sequence of rectifiable centrally symmetric curves lying in the surface of the unit ball such that their lengths approach 4 and such that they exhibit a certain convergence property or satisfy certain further location restrictions. For example, in Harrell-Karlovitz [2] it was shown that the requirement that the curves in the sequence all have a common point is again too strong. The following property is typical of the ones that are being presently studied. There exists a sequence $\{x_n\}$ in X such that every subspace generated by an infinite subsequence $\{x_{n'}\}$ again has girth 4. In a very natural manner this question can be restated in terms of tree properties.

3. SOME VARIATIONAL PROBLEMS IN HILBERT SPACE

Let X be a Hilbert space, either finite or infinite dimensional, with norm $\|\cdot\|$ and inner product $(\cdot, \cdot)$. Let $Q: X \to X$ be a compact, symmetric, positive definite linear operator. Let $\lambda_1 \geq \cdots \geq \lambda_n \geq \cdots$ be the eigenvalues of Q and $u_1, \ldots, u_n, \ldots$ the corresponding eigenvectors. Define the Rayleigh quotient in the usual manner

$$R(x) = (Qx, x)/(x, x).$$

The following variational characterizations of the eigenvalues are well-known:

A. $\quad \min_{S_n} \max_{x \perp S_n} R(x) = \lambda_{n+1},$

B. $\quad \max_{S_n} \min_{x \in S_n} R(x) = \lambda_n,$

where S_n ranges over all n-dimensional subspaces of X, and where $x \perp S_n$ means $(x, y) = 0$ for all $y \in S_n$.

If we let $S_n^* = \text{span}\{u_1, \ldots, u_n\}$ then it is easily verified that

$$\max_{x \perp S_n^*} R(x) = \lambda_{n+1} \quad \text{and} \quad \min_{x \in S_n^*} R(x) = \lambda_n.$$

Therefore S_n^* is said to be *extremal* for both variational characterizations A and B. We refer to S_n^* as the *classical* choice among all extremal subspaces. In general, the extremal

ON THE GEOMETRY AND STRUCTURE OF BANACH SPACES

subspaces of A and B are different (Remark 6 below). The determination of extremal subspaces, other than the classical choice, has had only limited study. Indeed, most discussions of A and B avoid this question altogether. Furthermore, we shall relate this to the study of extremal subspaces of certain Kolmogorov n-width problems, for which the uniqueness of the extremal subspace has been mistakenly assumed.

The set of all extremal subspaces of the variational characterization A was first given by Weinstein [18, 19]. Stenger [20] gave those of B. Our criteria are of an entirely different nature. We wish to stress the connection to the n-width problems as well as the geometric and approximation theoretic aspects.

The extremal subspaces of the n-width problems have not heretofore been characterized.

For each positive integer $n <$ dimension X, we define $q = q(n)$ and $m = m(n)$ by

$$\begin{aligned} &\lambda_q > \lambda_{q+1} = \cdots = \lambda_n, && \text{if such a } q \text{ exists,} \\ &&& \text{otherwise } q = 0, \\ &\lambda_{n+1} = \cdots = \lambda_{n+m} > \lambda_{n+m+1}, && \text{if such an } m \text{ exists,} \\ &&& \text{otherwise } m = \dim X - n. \end{aligned} \qquad (12)$$

THEOREM 9. ([21, Theorem 1]) If the n-dimensional subspace S_n is extremal for the variational characterization A (i.e., if $\max\{R(x): x \perp S_n\} = \lambda_{n+1}$) then

$$\begin{aligned} &S_n \perp u_{n+1}, \ldots, u_{n+m}, && \text{if } \lambda_n > \lambda_{n+1}, \\ &S_n \perp M, \text{ for some m-dimensional subspace } M \text{ of} \\ &\quad \text{span}\{u_{q+1}, \ldots, u_{n+m}\}, && \text{if } \lambda_n = \lambda_{n+1}, \end{aligned} \qquad (13)$$

where q and m are given by (12).

Conversely, if dimension $X > n + m$ and S_n satisfies

$$S_n \perp u_{n+1}, \ldots, u_{n+m},$$

and

$$\min\{\|u_k - y\|: y \in S_n\} \leq \varepsilon_k, \quad k = 1, \ldots, n, \tag{14}$$

with

$$\sum_{m=1}^{n} \varepsilon_k^2 \lambda_k \leq \lambda_{n+1} - \lambda_{n+m+1},$$

then S_n is extremal for the variational characterization A.

Remark 3. The sufficient condition (14) states that if S_n approximates the classical choice S_n^* closely enough then it is also extremal. Accordingly, all small perturbations of S_n^*, which do not involve the directions $u_{n+1}, \ldots, u_{n+m}$, preserve extremality.

Remark 4. By virtue of the necessary condition (13), the classical choice S_n^* plays a unique role among all extremal subspaces. For S_n^* is *known* to satisfy (13). On the other hand, to test an arbitrary subspace S_n for condition (13), we either need to know that $\lambda_n = \lambda_{n+1}$ or we need to have the vector u_{n+1} in hand. Thus if S_n is different from S_n^* we need to know either λ_{n+1} or u_{n+1} explicitly before we can test for extremality.

Remark 5. If $\lambda_n = \lambda_{n+1}$, then there are extremal subspaces, other than S_n^*, which are spanned by eigenvectors. Namely, subspaces of the form $S_n^{**} = \mathrm{span}\{u_1, \ldots, u_q, u_{j_1}, \ldots, u_{j_{n-q}}: q + 1 \leq j_1 < \cdots < j_{n-q} \leq n + m\}$. In a straightforward manner, the statement of condition (14) and its proof can be adjusted in order to describe extremality preserving perturbations of S_n^{**}.

THEOREM 10 ([21, Theorem 2]) If the n-dimensional subspace S_n is extremal for the variational characterization B (i.e., if $\min\{R(x): x \in S_n\} = \lambda_n$) then

ON THE GEOMETRY AND STRUCTURE OF BANACH SPACES

$$u_{q+1}, \ldots, u_n \in S_n, \text{ if } \lambda_n > \lambda_{n+1},$$

some $(n-q)$-dimensional subspace N of $\text{span}\{u_{q+1}, \ldots, u_{n+m}\}$

$$\text{belongs to } S_n, \text{ if } \lambda_n = \lambda_{n+1}, \quad (15)$$

where q and m are given by (12).

Conversely, if S_n satisfies

$$\min\{\|u_k - y\|: y \in S_n\} \leq \varepsilon_k, \quad k = 1, \ldots, n,$$

with

$$\varepsilon_k^2 \leq \min\{1/8r, (\lambda_r - \lambda_n)/8r\lambda_n\}, \quad k = 1, \ldots, r, \quad (16)$$

$$\varepsilon_k = 0, \quad k = r+1, \ldots, n,$$

for some r, $1 \leq r \leq q$, then S_n is extremal for the variational characterization B.

Remark 6. It follows readily from Theorems 9 and 10 that, in general, there exist subspaces which are extremal for the variational characterization A but not for B, and conversely.

Remark 7. The obvious analogs to Remarks 3, 4, and 5 hold for Theorem 10.

We now turn to some related Kolmogorov n-width problems. Since $Q: X \to X$ is symmetric and positive definite, we can define a new norm $\|\cdot\|_Q$, and corresponding inner product $(\cdot, \cdot)_Q$, on X by:

$$\|x\|_Q^2 = (x, x)_Q = (Qx, x).$$

In this norm, the unit ball of the original norm $\{x: \|x\| \leq 1\}$ can be expressed as the *ellipsoid*

$$K = \{x: \sum_i \lambda_i^{-1}(x, u_i/\lambda_i^{1/2})_Q^2 \leq 1\}.$$

We consider the problem of evaluating the (Kolmogorov) n-width of K, i.e.,

C. $\qquad w_n(K) = \min_{S_n} \max_{x \in K} \min_{y \in S_n} \|x - y\|_Q,$

where, as before, S_n ranges over all n-dimensional subspaces of X. The first minimum is clearly achieved; the maximum is achieved because K is compact in the $\|\cdot\|_Q$ norm; and the minimum over all S_n is shown below to be achieved. The number $w_n(K)$ is called (*Kolmogorov*) n-*width of* K. It measures the degree to which K can be approximated by the elements of a single n-dimensional subspace. A subspace S_n is said to be *extremal* for the n-width problem C if $\max\{\min\{\|x - y\|: y \in S_n\}, x \in K\} = w_n(K)$.

THEOREM 11   ([21, Theorem 3])   For each positive integer n, n < dimension X,

$$w_n(K) = (\lambda_{n+1})^{1/2}.$$

If the n-dimensional subspace S_n is extremal for the variational characterization A then it is also extremal for the n-width problem C, but not conversely in general.

The n-dimensional subspace S_n is extremal for the n-width problem C if and only if (13) and (17) are satisfied, where (17) is given by

$$\|x\|_Q \leq (\lambda_{n+1})^{1/2} \min\{\|x - y\|: y \in S_n\}, \qquad (17)$$

for all $x \in X$ such that $Qx \perp S_n$ and $x \perp u_{n+1}, \ldots, u_{n+m}$, if $\lambda_n > \lambda_{n+1}$; and for all x such that $Qx \perp S_n$ and $x \perp M$, where M is given by (13), if $\lambda_n = \lambda_{n+1}$.

ON THE GEOMETRY AND STRUCTURE OF BANACH SPACES

Remark 8. Since K is an ellipsoid, it follows readily from the geometry, or from (17), that $S_n^* = \text{span}\{u_1, \ldots, u_n\}$ is also extremal for the n-width problem C. In this context, we again refer to S_n^* as the *classical choice*. In most cases, as can be seen from (17) or even more easily from (14), nonclassical choices exist (for example, if $\lambda_n > \lambda_{n+1}$ and $n + 2 <$ dimension X). This is a matter that seems to be not well understood. In the original paper of Kolmogorov [22], wherein the notion of n-width is introduced, ellipsoids are considered and uniqueness is erroneously claimed. (See Example 4 below.) The same oversight appears in other discussions of n-widths, or the question of nonclassical choices is avoided altogether.

Remark 9. The combination of Theorems 9 and 11 leads to further criteria for extremality. For example, condition (14) is sufficient for a subspace to be extremal for the n-width problem C.

Remark 10. The nonuniqueness of the classical choice in n-width problems seems to be more useful than the nonuniqueness in the variational characterization of eigenvalues. In the numerical treatment of differential equations, it is often the case that computing the approximate solution in an extremal subspace minimizes the error between the approximate and actual solutions. In such cases, the nonuniqueness can be used to select an extremal subspace in a numerically advantageous manner.

In order to relate a width problem to the variational characterization B, we let

$$L = \{x: \sum_i \lambda_i^2 (x, u_i)^2 \leq 1\} = \{x: \|x\|_Q \leq 1\},$$

and we consider the problem of evaluating

D. $\quad w_{\text{co-dim. } n}(L) = \min_{S_n} \max_{x \in L} \min_{y \perp S_n} \|x - y\|,$

where S_n ranges over all n-dimensional subspaces of X. Since we are approximating with $y \perp S_n$, this can be considered as finding the width of L with subspaces of co-dimension n. As in C, the maxima and minima are readily shown to be achieved. The subspace S_n is said to be *extremal* for D if $\max\{\min\{\|x - y\|: y \perp S_n\} x \in L\} = w_{\text{co-dim. } n}(L)$.

THEOREM 12. ([21, Theorem 4]) For each positive integer n < dimension X,

$$w_{\text{co-dim. } n}(L) = (\lambda_n)^{1/2}.$$

If the n-dimensional subspace S_n is extremal for the width problem D then it is also extremal for the variational characterization B, but not conversely in general.

The n-dimensional subspace S_n is extremal for the width problem D if and only if (15) and (18) are satisfied, where (18) is given by

$$\|x\| \leq (1/\lambda_n)^{1/2} \min\{\|x - y\|: y \perp S_n\}, \qquad (18)$$

for all $x \in X$ such that $x \in S_n$ and $x \perp u_{q+1}, \ldots, u_n$, if $\lambda_n > \lambda_{n+1}$; and for all x such that $x \in S_n$ and $x \perp N$, where N is given by (15), if $\lambda_n = \lambda_{n+1}$.

Example 4. ([23]) One of the problems considered by Kolmogorov [22] is to find in $L^2[0, 1]$ the n-width of the set

$M = \{x: x, \frac{dx}{dt} \in L^2[0, 1], x \text{ absolutely continuous}, \|dx/dt\| \leq 1\}$.

He erroneously asserted that all extremal subspaces are unique. M can be expressed as M = K + L, where L is the span of the function $x_1(t) \equiv 1$, and where $K = \{x: \sum_k (<x, \sqrt{2} \cos k\pi> k\pi)^2 \leq 1, <x, x_1> = 0\}$. The latter is an ellipsoid of the form considered in the Theorem 11 above. Clearly if $n \geq 2$ then $w_n(M) = w_{n-1}(K)$

and the subspace spanned by x_1, cos πt, ..., cos$(n - 1)\pi t$ is the extremal for M. This is the classical choice. Moreover, if S_{n-1} is extremal for K then $L + S_{n-1}$ is extremal for M. Thus we can use Theorem 11 to find a nonclassical extremal subspace for M. For example, let u_1 and u_2 the piecewise linear, continuous real-valued functions defined on $[0, 1]$ with knots at $\{0, 1/6, 5/6, 1\}$. They are defined by: $u_1(0) = u_1(1/6) = 1$, $u_1(5/6) = u_1(1) = 0$, and $u_2(0) = u_2(1/6) = 0$, $u_2(5/6) = u_2(1) = 1$. It now readily follows that the subspace spanned by u_1 and u_2 is extremal for M. For $u_1 + u_2 = x_1$ and according to Theorem 11, $y_1 = u_1 - u_2$ spans an extremal subspace of dimension 1 for K. We note that y_1 is not in the direction of the major axis of K. It is a perturbation thereof which avoids the second major axis. The mistaken claim of uniqueness stems from the erroneous assumption that extremal subspaces of ellipsoids are necessarily spanned by their major axes.

REFERENCES

1. J. J. Schäffer, Inner diameter, perimeter, and girth of spheres, *Math. Ann.*, *173* (1967), 59-79.
2. R. E. Harrell and L. A. Karlovitz, Nonreflexivity and the girth of spheres, *Inequalities III*, Academic, New York (1972), 121-127.
3. R. E. Harrell and L. A. Karlovitz, Girths and flat Banach spaces, *Bull. Amer. Math. Soc.*, *76* (1970), 1288-1291.
4. R. E. Harrell and L. A. Karlovitz, The geometry of flat Banach spaces, *Trans. Amer. Math. Soc.*, *192* (1974), 209-218.
5. J. Lindenstrauss and C. Stegall, Examples of spaces which do not contain ℓ_1 and whose duals are non-separable (to appear).
6. R. E. Harrell and L. A. Karlovitz, Characterization of flat Banach spaces, *Pacific J. Math.*, (to appear).
7. P. Nyikos and J. J. Schäffer, Flat spaces of continuous functions, *Studia Math.*, *42* (1972), 221-229.

8. J. J. Schäffer, On the geometry of spheres in L-spaces, *Israel J. Math.*, 10 (1971), 114-120.

9. R. E. Huff, Dentability and the Radon-Nikodym property, *Duke Math. J.*, 41 (1974), 11-114.

10. I. M. Gelfand, Zur Theorie abstrakter Funktionen, *C. R. (Dokl.) Acad. Sci. URSS*, 17 (1937), 243-245.

11. R. C. James, Some self-dual properties of normed linear spaces, *Symposium on Infinite Dimensional Topology* Princeton University Press, Princeton, New Jersey, (1972), 159-175.

12. L. A. Karlovitz, On the duals of flat Banach spaces, *Math. Ann.*, 202 (1973), 245-250.

13. L. A. Karlovitz, On the subspace structure of flat Banach spaces, University of Maryland, Tech. Note.

14. J. Hagler, Embeddings of L^1 spaces into conjugate Banach spaces, Ph. D. Thesis (U. C. B., 1972).

15. L. A. Karlovitz, Construction of a Schauder basis for a class of nonreflexive Banach spaces, University of Maryland Tech. Note.

16. J. Schauder, Zur Theorie Abbildungen in Funktionalraumen, *Math. Z.*, 26 (1927), 47-65.

17. J. J. Schäffer and K. Sundaresan, Reflexivity and the girth of spheres, *Math. Ann.*, 184 (1970), 163-168.

18. A. Weinstein, Intermediate problems and the maximum-minimum theory of eigenvalues, *J. Math. Mech.*, 12 (1963), 235-246.

19. A. Weinstein, An invariant formulation of the new maximum-minimum theory of eigenvalues, *J. Math. Mech.*, 16 (1966), 213-218.

20. W. Stenger, On Poincaré's bounds for higher eigenvalues, *Bull. Amer. Math. Soc.*, 72 (1966), 715-718.

21. L. A. Karlovitz, Remarks on the variational characterization of eigenvalues and n-width problems, *J. Math. Anal. Appl.*, January 1976.

22. A. Kolmogorov, Über die beste Annäherung von Funktionen einer gegebenen Funktionenklasse, *Ann. Math.*, 37 (1936), 107-110.

23. L. A. Karlovitz, On a class of Kolmogorov n-width problems, *Atti del Accademia Nazionale dei Lincei*, LIII (1973), 241-245.

ANALYTIC TRANSFORMATION GROUPS

Wilhelm Kaup[†]

Departamento de Matemática
Universidade Federal do Rio de Janeiro
Rio de Janeiro, Brazil

In 1935 H. Cartan [1] proved the following theorem:

For every bounded domain $M \subset \mathbb{C}^n$ the group $G := \text{Aut}(M)$ of all biholomorphic automorphisms of M is a real Lie group of dimension $\leq n(n + 2)$ with respect to the compact open topology and the mapping $\Psi : G \times M \to M$ defined by $\Psi(g,x) = gx$ is real-analytic.

With Montel's theorem he showed that G is locally compact, and from this he derived that the set of all infinitesimal automorphisms (= complete holomorphic vector fields on M) is a real Lie algebra of finite dimension giving the Lie group structure on G. Later the result of Cartan was generalized (see, e.g., [2]):

For every connected differentiable (resp., real-analytic) manifold of finite dimension and every locally

---
[†] <u>Present address</u>: Mathematisches Institut der Universität Tübingen, Tübingen, Germany.

compact group G of differentiable (resp., analytic) isomorphisms of M which acts continuously on M (i.e., Ψ is continuous) there is a (unique) Lie group structure on G such that Ψ is differentiable (resp., analytic).

In particular, by a theorem of Bochner and Montgomery [3] the group

G = Aut(M) is a complex Lie group and Ψ is a holomorphic mapping if M is a compact complex manifold.

The natural objects in Complex Analysis to be studied are complex spaces (that are complex manifolds which "may have singular points"). Recently also complex spaces M of infinite dimension (= Banach analytic complex spaces) have become interesting, and for such an M the following question arises: Under which condition does there exist a Lie group structure (real or complex) on a given subgroup G $\subset$ Aut(M) making the mapping Ψ: G $\times$ M $\to$ M analytic?

In the following we shall give some results in connection with this problem. Therefore we have to work with a category which contains as objects complex spaces, real and complex Lie groups, and also more generally direct products G $\times$ M where G is a real Lie group and M is a complex space. For simplicity we take a very general category H of analytic objects (see Sec. 1). Then we have a unified theory for the real and the complex case, and furthermore there is no difficulty in carrying out certain constructions within H (e.g., the tangent bundle functor).

We shall not give proofs here (except for Theorem 2). The details have been given in a course at the Universidade Federal do Rio de Janeiro.

1. GENERALIZED ANALYTIC SPACES

Let E, F be complex Banach spaces, $U \subset E$ an open subset, and $H(U,F)$ the set of all holomorphic mappings $U \to F$ (see, e. g., [4]). Then the subset $B(U,F)$ of all bounded holomorphic mappings $U \to F$ is a complex Banach space with respect to the supremum norm. For every $A \subset E$, $B \subset F$ a map $f: A \to B$ is called *holomorphic* (or *analytic*) if to every $a \in A$ there is an open subset $U \subset E$ and $F \in H(U,F)$ such that $f|A \cap U = F|A \cap U$. Denote by $H(A,B)$ the set of all analytic mappings $A \to B$. The *generalized analytic spaces* are now obtained by patching together arbitrary subsets of complex Banach spaces via bianalytic transition functions in the usual way: $U = \{(U_i, \varphi_i, A_i): i \in I\}$ is called an *analytic atlas* on the Hausdorff topological space X if $\{U_i: i \in I\}$ is an open covering of X, every $\varphi_i: U_i \to A_i$ is a homeomorphism onto a subset A_i of some complex Banach space E_i and $\varphi_j \circ \varphi_i^{-1}: A_{ij} \to A_{ji}$ is analytic for every $i, j \in I$ such that $A_{ij} = \varphi_i(U_i \cap U_j) \neq \phi$. A maximal analytic atlas U on X is called an *analytic structure* on X and the pair $X = (X, U)$ is called a *generalized analytic space*. For two such spaces X,Y the notion of *analytic mapping* $X \to Y$ is defined in the obvious manner (take local charts), and we get the category H of generalized analytic spaces. This category admits finite direct products, and for every $X \in H$ and every subset $Y \subset X$ also Y is a generalized analytic space in a natural way. Note that $\mathbb{R}^2$ (as a subset of $\mathbb{C}^2$) and $\mathbb{C}$ are both generalized analytic spaces which are not bianalytically equivalent (for every analytic mapping $\mathbb{C} \to \mathbb{R}^2$ is a constant map).

For every complex Banach space E an $\mathbb{R}$-linear closed subspace $V \subset E$ is called an analytic *vector space*. The analytic structure on V is completely determined by the real Banach space V and the complex linear structure on the closed subspace $\tilde{V} := V \cap iV$. In particular, every real Banach space is an analytic vector space with $\tilde{V} = 0$ (put $E := V \oplus iV$ and $\|x + iy\| := \sup\{\|(\cos t)x + (\sin t)y\|: t \in \mathbb{R}\}$).

Let us call $X \in H$ *real* (resp., *complex*) if to every $x \in X$ there is a bianalytic mapping of a neighborhood of x onto a certain subset of some real Banach space (resp., if for every domain $U \subset X$ every analytic mapping $U \to \mathbb{R}$ is constant). Furthermore let us call $X \in H$ an *analytic space* if to every $x \in X$ there is a neighborhood U of x, an open subset W of an analytic vector space V, and a bianalytic mapping $\tau: U \to A$, where $A = f^{-1}(0) \subset W$ for some analytic mapping f from W into a Banach space E. If V always can be chosen in such a way that $A = V$, X is called an *analytic manifold*. In particular, we have the notion of real analytic space, complex analytic manifold, etc.

2. ANALYTIC TRANSFORMATION GROUPS

Let us call an analytic manifold G together with a group structure an *analytic group* if the mapping $G \times G \to G$ defined by $(x,y) \longmapsto xy^{-1}$ is analytic. If G is real (resp., complex) G is called a *Lie group* (resp., *complex Lie group*). A pair (G, Φ) is called an *analytic transformation group* on $X \in H$ if G is an analytic group and $\Phi: G \to \text{Aut}(X)$ is a homomorphism of groups such that the mapping

$$\Psi: G \times X \to X, \quad \Psi(g,x) = \Phi(g)(x),$$

is analytic. For example, for every complex Banach space E the group GL(E) of all continuous linear isomorphisms is a complex Lie group, and it is in a natural way an analytic transformation group on E.

Analytic transformation groups may be described infinitesimally by analytic vector fields, as in known for differentiable manifolds (see [5]): To every $X \in H$ there may be associated the Lie algebra $T(X)$ of analytic vector fields on X [these are analytic sections in the tangent bundle $T(X)$ over X].

ANALYTIC TRANSFORMATION GROUPS

In particular, we have for every analytic group G the Lie algebra $g \subset T(G)$ of all right-invariant vector fields, which is an analytic Lie algebra (i.e., g is an analytic vector space and the bracket $g \times g \to g$ is an analytic mapping). Then to every analytic transformation group (G,Φ) on X there is associated a Lie homomorphism $\varphi: g \to T(X)$ which is analytic in some sense and describes the action of G in a neighborhood of $e \in G$ completely. By this the study of analytic transformation groups is reduced to that of a linear problem. Therefore the following problem is interesting:

Let G be a topological group and $\Phi: G \to \text{Aut}(X)$, $X \in H$, a homomorphism such that $\Psi: G \times X \to X$ is continuous [then (G,Φ) is called a *continuous transformation group on* X]. Does there exist an analytic group structure on G such that (G,Φ) is an analytic transformation group?

To get results we have to make assumptions. It can be shown that every analytic transformation group (G,Φ) satisfies the following stronger continuity condition:

DEFINITION. Let G be a topological group and $\Phi: G \to \text{Aut}(X)$, $X \in H$, a homomorphism of groups. Then (G,Φ) is called a *strongly continuous transformation group* if for every $x \in X$ there are a neighborhood Q of $e \in G$, an open subset U of some complex Banach space E, a continuous mapping $f: Q \to B(U,E)$, and a neighborhood A of $x \in X$ which may be realized as a subset $A \subset U$ such that the following is true: For every $g \in Q$ and $a \in A$ with $ga = \Phi(g)a \in A$ we have $ga = f(g)a$.

Obviously "strongly continuous" means in particular that for Q and A suitable the neighborhood A may be realized as a subset of E in such a way that the germ of $\Phi(g)$ at $x \in A$ is the restriction of a holomorphic mapping defined in a neighborhood of $x \in E$ with radius of convergence bounded away from 0 on Q.

Let us call $X \in H$ *irreducible* if for every analytic function $f: X \to \mathbb{C}$ the following is true: $f|U = 0$ for some open subset $U \neq \emptyset$ of X always implies $f = 0$. For instance, every connected analytic manifold is irreducible in this sense and we have [5]:

THEOREM 1. Suppose $X \in H$ is irreducible and (G,Φ) is a strongly continuous transformation group on X with G locally compact. Then if (G,Φ) is effective (i.e., Φ is injective), there is a unique analytic group structure on G such that the following is true:

(i) (G,Φ) is an analytic transformation group on X.

(ii) For every analytic transformation group $(\tilde{G},\tilde{\Phi})$ on X and every continuous homomorphism α such that the following diagram commutes

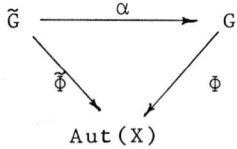

the mapping α is analytic.

It can be seen easily that for every compact complex manifold M the complex Lie group structure on $G = \mathrm{Aut}(M)$ given by the theorem of Bochner and Montgomery as well as the real Lie group structure on $G = \mathrm{Aut}(M)$ for a bounded domain $M \subset \mathbb{C}^n$ are the structures characterized by Theorem 1. In particular, $\mathrm{Aut}(M)$ for a bounded domain $M \subset \mathbb{C}^n$ is a complex Lie group only in the case that $\mathrm{Aut}(M)$ is discrete.

As an easy consequence of Theorem 1 we obtain the following well-known result (note that every subgroup H of an analytic group G acts strongly continuous on G by left translations):

COROLLARY. Every locally compact subgroup H of an analytic group G is an analytic group.

The condition that $\Psi: G \times X \to X$ be continuous [that is, that (G,Φ) is a continuous transformation group] is quite natural; it is satisfied, for instance, if X is locally compact and $G \subset \text{Aut}(X)$ is endowed with the compact open topology. In contrast to this, "strongly continuous" is more technical in nature, and the question is whether it may be substituted by the weaker condition "continuous transformation group" in Theorem 1.

A first result in this direction is the following:

THEOREM 2. Suppose (G,Φ) is a continuous transformation group on $X \in \mathcal{H}$. Then (G,Φ) is strongly continuous if

(i) G is locally compact and

(ii) X is locally compact and has finite embedding dimension (i.e., every $x \in X$ admits a compact neighborhood A which can be realized as a subset of some $\mathbb{C}^n$).

Proof. Fix a point $x \in X$ and realize a certain compact neighborhood A of x as a subset $A \subset \mathbb{C}^n$. Then there exist a compact neighborhood Q of $e \in G$, a compact neighborhood V of $x \in \mathbb{C}^n$, and a fundamental system $\{U_\nu : \nu \in \mathbb{N}\}$ of Stein neighborhoods of $x \in \mathbb{C}^n$ such that for $A_\nu := A \cap U_\nu$ we have $Q(A_\nu) \subset V \cap A$, $\nu \in \mathbb{N}$. Denote by $B_\nu \subset U_\nu$ the smallest analytic subset with $A_\nu \subset B_\nu$. Then every B_ν is a complex space and $F_\nu := H(B_\nu, \mathbb{C}^n)$ is a Fréchet-Montel space. The restriction mapping $\rho: F_\nu \to H(A_\nu, \mathbb{C}^n)$ is continuous, linear, and injective (every function space here is endowed with the compact open topology).

By Montel's theorem

$$K_\nu := \{g \in F_\nu : g(B_\nu) \subset V,\ g(A_\nu) \subset A\}$$

is compact. The restriction mapping $\varphi: Q \to H(A_\nu, \mathbb{C}^n)$ also is continuous, i.e.,

$$Q_\nu := \varphi^{-1}(\rho(K_\nu)) \quad \text{is closed in } Q$$

for every ν. Obviously we have

$$Q = \bigcup_{\nu \in \mathbb{N}} Q_\nu.$$

By a category argument (Q is a Baire space), therefore, some Q_ν has an inner point h. Using the translation by h^{-1} we may assume $Q = Q_\nu$ for some fixed $\nu \in \mathbb{N}$. Because K_ν is compact

$$\rho: K_\nu \to \varphi(Q_\nu)$$

is a homeomorphism, i.e., there is a continuous mapping τ (namely $\tau = \rho^{-1} \circ \varphi$) such that

$$\begin{array}{c} \xrightarrow{\tau} H(B_\nu, \mathbb{C}^n) \\ Q \downarrow \rho \\ \xrightarrow{\varphi} H(A_\nu, \mathbb{C}^n) \end{array}$$

is commutative. Because U_ν is Stein the restriction mapping $\sigma: H(U_\nu, \mathbb{C}^n) \to H(B_\nu, \mathbb{C}^n)$ is a continuous, linear, and *surjective* mapping of Fréchet spaces. By a theorem of Cartan [6], therefore, there exists a continuous λ such that

$$\begin{array}{c} \xrightarrow{\lambda} H(U_\nu, \mathbb{C}^n) \\ Q \downarrow \sigma \\ \xrightarrow{\tau} H(B_\nu, \mathbb{C}^n) \end{array}$$

is commutative. Then take a relatively compact open neighborhood $U \subset U_\nu$ of $x \in \mathbb{C}^n$ and define

$$f: Q \to B(U, \mathbb{C}^n)$$

by $f(g) = \lambda(g)|U$.   Q.e.d.

REFERENCES

1. H. Cartan, Sur les groupes de transformations analytiques, *Actualités Sci. Ind.* 198, (1935).
2. D. Montgomery and L. Zippin, *Topological Transformation Groups*, Wiley-Interscience, New York, 1965.
3. S. Bochner and D. Montgomery, Groups of differentiable and real or complex analytic transformations, Ann. Math., 46 (1945), 685-694.
4. L. Nachbin, Topology on spaces of holomorphic mappings, *Ergebnisse der Mathematik*, 45, Springer, Berlin, 1969.
5. W. Kaup, Transformation groups in complex analysis, in preparation for *Ergebnisse der Mathematik*, Springer, Berlin.
6. H. Cartan, Espaces fibrés analytiques, Symp. *International de Topología Algebraica*, Mexico, 1968, p. 119.

ON LOCALLY CONVEX SPACES WITH THE MONTEL PROPERTY

Mário C. Matos

Departamento de Matemática do IMECC
Universidade Estadual de Campinas
Campinas, São Paulo, Brazil

This chapter discusses some results concerning those locally convex spaces to which the Montel theorem is naturally generalized.

If E and F are complex locally convex spaces and U is a nonvoid open subset of E, $H(U; F)$ denotes the vector space of all holomorphic mappings from U into F. Two locally convex topologies will be considered in $H(U; F)$: (i) the topology τ_0 generated by the set of seminorms $\{p_{K,\beta}; K \subset U, K \text{ compact, and } \beta \text{ a continuous seminor in } F\}$, and (ii) the topology τ_{0f} generated by $\{p_{K,\beta}; K \subset U \text{ compact and finite dimensional and } \beta \text{ a continuous seminorm in } F\}$. The seminorm $p_{K,\beta}$ is defined by $p_{K,\beta}(f) = \sup\{\beta[f(t)]; t \in K\}$ for each f in $H(U; F)$. $C(U; F)$ denotes the set of all τ_0-relatively compact subsets of $H(U; F)$. $B(U; F)$ is the set of all τ_0-bounded subsets X of $H(U; F)$ such that $X(u) = \{f(u); f \in X\}$ is relatively compact in F for every u in U. $B_f(U; F)$ denotes the set of all τ_{0f}-bounded subsets X of $H(U; F)$ such that $X(u)$ is relatively compact in F for each u of U. It is clear that $C(U; F) \subset B(U; F) \subset B_f(U; F)$ and that the classical

Montel theorem establishes that for $n \in \mathbb{N}$, $E = \mathbb{C}^n$, $C(U; F) = B(U; F) = B_f(U; F)$ for each U and F. This motivates the following definitions:

DEFINITION 1. E satisfies the first Montel property if $C(U; F) = B_f(U; F)$ for all U and F.

DEFINITION 2. E satisfies the second Montel property if $C(U; F) = B(U; F)$ for all U and F.

It obvious that every locally convex space satisfying the first Montel property satisfies the second and that for $E = \mathbb{C}^n$, $n \in \mathbb{N}$, E satisfies both Montel properties. There are locally convex spaces which do not satisfy the second Montel property:

Example 1. Let I be a set whose power is at least equal to that of the continuum. The topological direct sum $E = \mathbb{C}^{(I)}$ of $\mathbb{C}$ indexed by I does not satisfy the second Montel property. In fact, there is a homogeneous polynomial p from E into $\mathbb{C}$ of degree 2 which is not continuous. Let $c_{ij} \in \mathbb{C}$ ($i, j \in I$) be such that $p(x) = \Sigma_{i,j \in I} c_{ij} x_i x_j$ for every $x = (x_i)_{i \in I} \in E$. If $J \subset I$ is finite, define $p_J: E \to \mathbb{C}$ by $p_J(x) = \Sigma_{i,j \in J} c_{ij} x_i x_j$ for each $x = (x_i)_{i \in I} \in E$. Each p_J is a continuous homogeneous polynomial of degree 2 and the set $X = \{p_J; J \subset I, J \text{ finite}\} \in B(E; \mathbb{C})$. However, X is not τ_0-relatively compact because any subnet of $\{p_J\}_{J \subset I}$, J finite, converges pointwise to p, which is not continuous.

In order to give more examples of locally convex spaces satisfying either the first or the second Montel property we recall the following definitions from [1] and [2].

DEFINITION 3. E is holomorphically barreled if, for every U and F, each $X \subset H(U; F)$ which is τ_{0f}-bounded is amply bounded, that is $\{\beta \circ f; f \in X\}$ is locally bounded for all continuous seminorms β in F.

DEFINITION 4. E is holomorphically infrabarreled if, for every U and F, each $X \subset H(U; F)$ which is τ_0-bounded is amply bounded.

It is proved in [1] that Silva spaces, Baire spaces, and arbitrary products of Fréchet spaces are holomorphically barreled (hence holomorphically infrabarreled) and arbitrary products of semimetrizable spaces are holomorphically infrabarreled. The following proposition is proved via Ascoli's Theorem:

PROPOSITION 1. Every holomorphically barreled space satisfies the first Montel property and every holomorphically infrabarreled space satisfies the second Montel property.

It is not difficult to see that the space c_{00} of the finite sequences of complex numbers normed by the ℓ^2 norm satisfies the second but not the first Montel property.

A modification of the proof of Proposition 3 of [1] proves the following result:

PROPOSITION 2. Let E be a topological direct sum of a sequence $(E_m)_{m \in \mathbb{N}}$ of normed (respectively, Banach) spaces. In order that E satisfy the second (respectively, first) Montel property it is sufficient that either each E_m is finite dimensional or $E_m = \{0\}$ except for a finite number of m's. This condition is necessary for E to be such that $C(E; \mathbb{C}) = B(E; \mathbb{C})$ [respectively, $C(E; \mathbb{C}) = B_f(E; \mathbb{C})$].

This proposition and Proposition 3 of [1] allow us to write:

PROPOSITION 3. Let E be a topological direct sum of a sequence of normed (respectively, Banach) spaces. Then E satisfies the second (respectively, first) Montel property if, and only if, E is holomorphically infrabarreled (respectively, barreled).

Besides being a natural question, the preceding proposition motivates the following problem which is open:

PROBLEM. Is every locally convex space which satisfies the first (respectively, the second) Montel property holomorphically barreled (respectively, infrabarreled)?

Example 2. Proposition 2 shows that $X \times Y$ does not satisfy the second Montel property if X is an infinite-dimensional normed or Banach space and $Y = \mathbb{C}^{(\mathbb{N})}$, which is a Silva space. This proves that the Montel properties do not remain true for Cartesian products of spaces having them.

DEFINITION 5 (Dineen). E is a Zorn space if for every U, F, and $f: U \to F$ finitely holomorphic in U, the set of points where f is continuous is open and closed in U.

DEFINITION 6. E is a weak Zorn space if for every U, F, and $f: U \to F$ hipoholomorphic in U (i.e., finitely holomorphic in U and continuous over the compact subsets of U), the set of points of U where f is continuous is open and closed.

The following interesting result is proved:

PROPOSITION 4. If E satisfies the first (respectively, second) Montel property then E is a Zorn (respectively, weak Zorn) space.

The next example shows that the converse of Proposition 4 does not hold.

Example 3. Let $E = \mathbb{C}^{\mathbb{N}} \times \mathbb{C}^{(\mathbb{N})}$, which is a Zorn space by a result of Hirschowitz. The mapping $f: E \to \mathbb{C}$ defined by $f(x, y) = \sum_{k=0}^{\infty} x_k y_k$ for $x = (x_k)_{k \in \mathbb{N}} \in \mathbb{C}^{\mathbb{N}}$ and $y = (y_k)_{k \in \mathbb{N}} \in \mathbb{C}^{(\mathbb{N})}$ is not continuous on E. For each $n \in \mathbb{N}$, we consider

$$f_n(x, y) = \sum_{k=0}^{n} x_k y_k$$

for $x = (x_k)_{k \in \mathbb{N}} \in \mathbb{C}^{\mathbb{N}}$ and $y = (y_k)_{k \in \mathbb{N}} \in \mathbb{C}^{(\mathbb{N})}$. It is easy to see that $X = \{f_n; n \in \mathbb{N}\} \in \mathcal{B}(E; \mathbb{C})$. But $X \notin C(E; \mathbb{C})$ since every subset of $(f_n)_{n \in \mathbb{N}}$ converges pointwise to f, which is not continuous in E.

DEFINITION 7. E satisfies the first (respectively, second) Montel property locally if, for every U and each $u \in U$, there is an open neighborhood V of u in U such that $C(V; F) = \mathcal{B}_f(V; F)$ [respectively, $C(V; F) = \mathcal{B}(V; F)$] for all F.

It is clear that spaces satisfying a Montel property satisfy it locally. Up to now all the known spaces which satisfy a Montel property locally are also spaces which satisfy a Montel property.

Proposition 4 holds for E satisfying the Montel property locally.

REFERENCES

1. J. Barroso, M. C. Matos, and L. Nachbin, On bounded sets of holomorphic mappings, in *Proc. Infinite Dimensional Holomorphy*, Univ. Kentucky, 1973; *Lecture Notes in Mathematics*, No. 364, Springer-Verlag, New York, 1974.

2. L. Nachbin, A glimpse at Infinite Dimensional Holomorphy, in *Proc. Infinite Dimensional Holomorphy*, Univ. Kentucky, 1973; *Lecture Notes in Mathematics*, No. 364, Springer-Verlag, New York, 1974.

SOME HOLOMORPHICALLY SIGNIFICANT
PROPERTIES OF LOCALLY CONVEX SPACES

Leopoldo Nachbin

Instituto de Matemática
Universidade Federal do Rio de Janeiro
Rio de Janeiro, Brazil

1. INTRODUCTION

Functional analysts were led to classify locally convex spaces according to noteworthy properties from the linear viewpoint. There resulted the classes of bornological, or barreled, or infrabarreled, or Mackey, spaces, among others (see the references to books on locally convex spaces at the end of this article). No matter how important they are from the strictly linear aspect, now they look too broad to be suitable for applications.

There is a fruitful interplay between the linear theory, that is, the study of the space $L(E; F)$ of all continuous linear mappings of E into F, and the holomorphic theory, that is, the investigation of the space $H(U; F)$ of all holomorphic mappings of the nonvoid open subset U of E into F, where E and F denote complex locally convex spaces. It is due, although not exclusively, to the fact that

$$L(E; F) \subset H(E; F),$$

or even, more generally,

$$L(E; F) \mid U \subset H(U; F).$$

For instance, certain studies of $L(E; F)$ suggest analogous, although modified, considerations concerning $H(U; F)$. As a result, for example, we have the important classes of holomorphically bornological, or holomorphically barreled, or holomorphically infrabarreled or holomorphically Mackey, spaces, among others [13, 22]. These new classes are more precise and limited, thus being more apt to fill the needs of current applications of functional analysis. In particular, although the notion of a Mackey space is usually agreed upon to be too general, that of a holomorphically Mackey space is less so; it corresponds to the very desirable condition that holomorphy be identical to weak holomorphy.

Let us introduce the following abbreviations for properties of a complex locally convex space: B = Baire, S = Silva, sm = semimetrizable, hba = holomorphically barreled, hbo = holomorphically bornological, hib = holomorphically infrabarreled, hM = holomorphically Mackey. We then have the following implications for the named properties of a complex locally convex space:

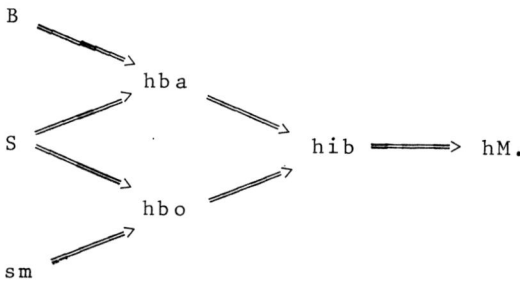

These results correspond to known ones, when we replace $H(U; F)$ by $L(E; F)$.

The program of pushing ahead and clarifying analogies, and the interplay, between $L(E; F)$ and $H(U; F)$, that is, the linear and holomorphic aspects, is timely and important from the viewpoint of applications of functional analysis. Such a program is being carried on at present in several directions; we quote particularly [19] and [15].

The inspiration for studying infinite dimensional holomorphy came from linear functional analysis and from several complex variables.

However, it was found that there are problems in infinite dimensional holomorphy which do not arise in either linear functional analysis or several complex variables. There are classes of complex locally convex spaces which are not pinned down in the linear theory, but which become of importance when we associate to them the spaces of holomorphic mappings.

We shall adopt here and use freely the notation and terminology presented in [22] and [21] (see also the references to books on infinite dimensional holomorphy at the end of the article). Unless the contrary is stated explicitly, E and F will always denote complex locally convex spaces, and U will be a nonvoid open subset of E.

For the major aspects of the present text, we quote particularly [13], [14], [16], [20], and [22].

2. HOLOMORPHICALLY BORNOLOGICAL SPACES

The history of the concept of bornological spaces is the following.

If E and F are normed spaces, then a linear mapping f: E → F is continuous if and only if it is bounded in the sense that it maps bounded subsets of E into bounded subsets of F. Now, if E and F are locally convex spaces, then whenever f is continuous it certainly maps bounded subsets of E into bounded subsets of F; however, the converse is false, that is, it may occur that f maps every bounded subset of E into a bounded subset of F without being continuous. The simple examples showing that such a converse is not true are due to the fact that E does not behave properly, in the sense that its topology is not large enough with respect to its collection of bounded subsets. This leads us to put our finger on E, and set the following classical definition.

DEFINITION 1. E is a *bornological space* if, for every F, we have that each mapping f: E → F belongs to $L(E; F)$ if (and always only if) f is linear, and f is bounded on every bounded subset of E.

Remark 1. In Definition 1, use is made of an arbitrary F and of $L(E; F)$. As suitable examples show, it is not enough to use only F = $\mathbb{C}$ in Definition 1. However, it can be seen that the following definition, which involves only E, is equivalent to Definition 1; namely, E is a bornological space if each seminorm on E which is bounded on every bounded subset of E must be continuous.

Remark 2. The simplest criterion is the following: If E is semimetrizable, then it is bornological. In particular, if E is normed, as in our starting considerations, then it is bornological.

Continuous linear mappings and bounded subsets get along very well, in the sense that each $f \in L(E; F)$ is bounded on every bounded subset of E. However, the same is not true in the infinite dimensional holomorphic case, in the following sense: it may occur that $f \in H(E; F)$ will be unbounded on some bounded subset of E.

Example 1. Let $E = c_0$ be the Banach space of all sequences of complex numbers tending to zero, endowed with its supremum norm, and $F = \mathbb{C}$. If we define

$$f(x) = \sum_{m=1}^{\infty} (x_m)^m$$

for every

$$x = (x_1, \ldots, x_m, \ldots) \in E,$$

then $f \in H(E; F)$. For every $m = 1, 2, \ldots$, if we consider a point $x \in E$ having m of its coordinates equal to 1, and all the remaining coordinates equal to 0, then $\|x\| = 1$ and $f(x) = m$. It follows that f is unbounded on the closed ball in E of center at 0 and radius equal to 1.

For additional valuable information on aspects related to Example 1, we refer to the Appendix at the end of the article.

Since holomorphic mappings and bounded subsets do not get along very well, in the sense that we just described, it is fortunate for our present considerations that Definition 1 is equivalent to the following Definition 2, which is not usually pointed out in the current expositions of topological vector spaces.

DEFINITION 2. E is a *bornological space* if, for every F, we have that each mapping $f: E \to F$ belongs to $L(E; F)$ if (and always only if) f is linear, and f is bounded on every compact subset of E.

Remark 3. As in Remark 1, we point out that it can be seen that the following definition, which involves only E, is equivalent to Definition 2; namely, E is bornological if each seminorm on E which is bounded on every compact subset of E must be continuous.

Holomorphic mappings and compact subsets get along very well, in the sense that each f ∈ H(U; F) is continuous and thus bounded on every compact subset of U. As a result, we shall pick up Definition 2 as the motivation for the following Definition 3.

DEFINITION 3. E is a *holomorphically bornological space* if, for every U and every F, we have that each mapping f: U → F belongs to H(U; F) if (and always only if) f is algebraically holomorphic, and f is bounded on every compact subset of U.

We recall that a mapping f: U → F is said to be algebraically holomorphic if, for every finite dimensional vector subspace S of E meeting U, we have that the restriction mapping

$$f \mid (U \cap S): U \cap S \to F$$

is holomorphic when S is endowed with its largest (= separated) locally convex topology; this concept does not depend on the given topology on E.

Remark 4. By comparing Definitions 2 and 3, and since linear mappings are algebraically holomorphic, it is clear that a holomorphically bornological space is also a bornological space.

PROPOSITION 1. A semimetrizable space E is holomorphically bornological.

Remark 5. Proposition 1 is an improvement of the known fact that a semimetrizable space E is bornological.

PROPERTIES OF LOCALLY CONVEX SPACES

PROPOSITION 2. A Silva space E is holomorphically bornological.

We recall that E is said to be a Silva space if there is a sequence E_m of Banach spaces, where E_m is a vector subspace of E and $E_m \subseteq E_{m+1}$, the inclusion mapping $E_m \hookrightarrow E_{m+1}$ being compact for every $m = 1, 2, \ldots$; moreover, it is assumed that

$$E = \bigcup_m E_m,$$

and that E carries the inductive limit topology. We also recall that Silva spaces are essentially the same thing as strong dual spaces of Fréchet-Schwartz spaces. More explicitly, the strong dual space of a Fréchet-Schwartz space is a Silva space; the strong dual space of a Silva space is a Fréchet-Schwartz space; and both Silva spaces and Fréchet-Schwartz spaces are reflexive.

The proof of Proposition 2 rests on the following Lemma 1.

LEMMA 1. Let E be a complex vector space, E_m be a locally convex space, $\rho_m: E_m \to E$ be a linear mapping, and $\sigma_m: E_m \to E_{m+1}$ be a compact linear mapping such that $\rho_m = \rho_{m+1} \circ \sigma_m$ for $m = 1, 2, \ldots$. Assume that

$$E = \bigcup_m \rho_m(E_m)$$

and endow E with the inductive limit topology. Let $U \subset E$ be open. Put $U_m = \rho_m^{-1}(U)$ for $m = 1, 2, \ldots$, and assume that U_0 is nonvoid. If F is a complex locally convex space and $f: U \to F$, then $f \in H(U; F)$ if and only if $f \circ \rho_m \in H(U_m; F)$ for every $m = 1, 2, \ldots$.

Remark 6. Lemma 1 is a reminiscence of the known fact that, if E_i ($i \in I$) is any family of locally convex spaces, E

is a vector space, $\rho_i: E_i \to E$ ($i \in I$) is a linear mapping, E is
endowed with the inductive limit topology, and F is a locally
convex space, then a linear mapping $f: E \to F$ is continuous if
and only if $f \circ \rho_i: E_i \to F$ is continuous for every $i \in I$. In
this connection, it is to be noticed that Lemma 1 may break down
in the absence of the denumerability, or compactness, conditions.

Remark 7. Proposition 2 is a reminiscence of the known
fact that any inductive limit of bornological spaces is
bornological. In this connection, it is to be noticed that a
denumerable inductive limit of holomorphically bornological
spaces may fail to be holomorphically bornological.

PROPOSITION 3. If E is holomorphically bornological, then $H(U; F)$
is complete for the compact-open topology, provided F is complete.

Remark 8. It is known that, if E is bornological and F is
complete, then $L(E; F)$ is complete both with respect to its
strong and the compact-open topology. Proposition 3 corresponds
to the second half of this remark.

Example 2. We now show that compactness is needed in Lemma
1, even in the presence of denumerability. Let X_0 be an infinite
dimensional metrizable complex locally convex space. Choose
$g_m \in H(X_0; \mathbb{C})$ for $m = 1, 2, \ldots$ so that, given any neighborhood
V of 0 in X, there is some g_m which is unbounded on V. It is
known that such a sequence exists if X_0 is not normable; in
particular, it is clear that such a sequence exists if $X_0 = \mathbb{C}^{\mathbb{N}}$,
for we may take g_m as the various projections of X_0 onto $\mathbb{C}$. It
also has been known for quite some time that such a sequence
exists if X_0 is normable, and separable or reflexive; actually,
it follows from the Josefson-Nissenzveig theorem (see Appendix)
that such a sequence exists if X_0 is just normable. Anyway, set

$X_m = \mathbb{C}$ for $m = 1, 2, \ldots$. Consider the topological direct sum

$$E = \sum_{m=0}^{\infty} X_m$$

and define $f: E \to \mathbb{C}$ by

$$f(x) = \sum_{m=1}^{\infty} g_m(x_0) \, x_m$$

if $x = (x_m) \in E$. Letting

$$E_m = X_0 \oplus \cdots \oplus X_m \oplus 0 \oplus \cdots,$$

then $f \mid E_m \in H(E_m; \mathbb{C})$ for $m = 0, 1, \ldots$. Notice that each E_m is metrizable, and that each E_m is normable if X_0 is normable. However, f is unbounded at 0; hence $f \notin H(E; \mathbb{C})$. This shows that Lemma 1 breaks down if the σ_m are assumed to be linear and only continuous, but not compact.

Example 3. We now show that denumerability is needed in Lemma 1, even in the presence of compactness. Consider a complex vector space E whose algebraic dimension is at least equal to the power of the continuum, endowed with its largest locally convex topology, which is the inductive limit of its finite dimensional vector subspaces. It is known that there is a homogeneous polynomial of degree two $p: E \to \mathbb{C}$ which is not continuous. For every finite dimensional vector subspace $X \subset E$, it is clear that $p \mid X \in H(X; \mathbb{C})$; however, $p \notin H(E; \mathbb{C})$. This shows that Lemma 1 breaks down if we deal with a nondenumerable family.

Example 4. Consider the Cartesian product

$$E = \mathbb{C}^{\mathbb{N}} \times \mathbb{C}^{(\mathbb{N})}$$

of an infinite denumerable Cartesian product of $\mathbb{C}$, by an infinite denumerable direct sum of $\mathbb{C}$, endowed with their natural topologies. Let $f: E \to \mathbb{C}$ be the bilinear form defined by

$$f(x, y) = \sum_{m=0}^{\infty} x_m y_m$$

if

$$x = (x_m) \in \mathbb{C}^{\mathbb{N}}, \quad y = (y_m) \in \mathbb{C}^{(\mathbb{N})};$$

thus f is algebraically holomorphic, since it is a homogeneous polynomial of degree two on E. It is clear that f is bounded on the compact subsets of E. However, f is not continuous, hence not holomorphic on E. It follows that E is not holomorphically bornological. However, E is bornological as the Cartesian product of two bornological spaces. Hence, a bornological space may not be holomorphically bornological. We also see that the Cartesian product of two holomorphically bornological spaces may fail to be holomorphically bornological. This example also shows that a denumerable direct sum, or a denumerable inductive limit, of holomorphically bornological space does not have to be holomorphically bornological, since

$$\mathbb{C}^{\mathbb{N}} \oplus \mathbb{C} \oplus \cdots$$

and E are isomorphic and homeomorphic. Actually, $H(E; \mathbb{C})$ is not complete, or even sequentially complete, for the compact-open topology. To see this, it is enough to introduce the truncated bilinear form $f_k: E \to \mathbb{C}$ defined by

$$f_k(x, y) = \sum_{m=0}^{k} x_m y_m;$$

then $f_k \in H(E; \mathbb{C})$ for $k = 0, 1, \ldots$. Since $f_k \to f$ uniformly on every compact subset of E as $k \to \infty$, but $f \notin H(E; \mathbb{C})$, we conclude that $H(E; \mathbb{C})$ is not sequentially complete for the compact-open topology. More generally, if X is a complex locally convex space which is not seminormable, and X' is its strong dual space, then

$$E = X \times X',$$

PROPERTIES OF LOCALLY CONVEX SPACES

with the Cartesian product topology, is not holomorphically bornological. As a matter of fact, the natural bilinear form

$$(x, y) \in E \mapsto y(x) \in \mathbb{C},$$

establishing the duality between X and X', is algebraically holomorphic, since it is a homogeneous polynomial of degree two on E. It is clear that it is bounded on the bounded subsets of E. However, it is not continuous, hence not holomorphic on E.

Example 5. We now show that it is not enough to use only $F = \mathbb{C}$ in Definition 3. Let E be a complex vector space. Assume that S and T are two locally convex topologies on E such that:

1. $S \subset T$ and $S \neq T$;

2. S and T have the same compact subsets of E, hence the same bounded subsets of E;

3. for every nonvoid subset $U \subset E$ that is open for S, hence for T, $H(U; \mathbb{C})$ is the same regardless of the fact that we endow E with S or T;

4. E is holomorphically bornological when it is endowed with T.

Then, we claim that E endowed with S is not bornological, hence not holomorphically bornological; however, for every nonvoid subset $U \subset E$ that is open for S, each function $f: U \to \mathbb{C}$ which is algebraically holomorphic, and is bounded on every compact subset of U, must be holomorphic, E being endowed with S. In fact, E endowed with S is not bornological, since the identity mapping

$$I: (E, S) \to (E, T)$$

is linear, and it maps bounded subsets into bounded subsets, but it is not continuous. Now let $U \subset E$ be nonvoid and open for S, and let the function $f: U \to \mathbb{C}$ be algebraically holomorphic and bounded on every subset of U which is compact for S. Since U is open for T, and the function $f: U \to \mathbb{C}$ is algebraically holomorphic and bounded on every subset of U which is compact for T, then f is holomorphic if E is endowed with T, since (E, T) is holomorphically bornological. It follows that f is holomorphic if E is endowed with S. An example of this situation is the following. Take a nondenumerable set and set $E = c_0(I)$, that is, the vector space of all complex valued functions on I tending to zero at infinity. Let T be the topology on E defined by the full supremum norm

$$x = (x_i) \in E \longmapsto \sup_{i \in I} |x_i| \in \mathbb{R};$$

whereas let S be the topology on E defined by the family of the denumerable supremum seminorms

$$x = (x_i) \in E \longmapsto \sup_{i \in J} |x_i| \in \mathbb{R},$$

for all denumerable $J \subset I$. Then, all the above four conditions can be checked; the third condition was pointed out by Josefson [16].

Remark 9. Example 5 also shows that it is not enough to use only $F = \mathbb{C}$ in Definition 1. However, the simpler classical example for this fact is the following. Let E be a bornological space which is not a weak space. Call T the given topology on E, and set $S = \sigma(E, E')$ to be the weak topology defined on E by E'. Every linear form on E that is bounded on the subsets of E which are bounded for S is also bounded on the subsets of E which are bounded for T. Then, it is continuous for T. It follows that it is continuous for S. However, E endowed with S is not bornological, since the identity mapping

$$I: (E, S) \to (E, T)$$

is linear, and it maps bounded subsets into bounded subsets, but it is not continuous.

3. HOLOMORPHICALLY BARRELED SPACES AND HOLOMORPHICALLY INFRABARRELED SPACES

The history of the concept of barreled spaces is the following.

If E is a Banach space and F is a normed space, then the classical Banach-Steinhaus theorem says that a collection $X \subset L(E; F)$ is bounded, that is,

$$\sup_{f \in X} \|f\| < +\infty,$$

if and only if it is pointwise bounded on E, that is,

$$\sup_{f \in X} \|f(x)\| < +\infty$$

for every $x \in E$. More generally, if E is a Fréchet space and F is a locally convex space, then the classical Banach-Steinhaus theorem says that a collection $X \subset L(E; F)$ is equicontinuous if and only if it is pointwise bounded on E, that is, the subset

$$X(x) = \{f(x);\ f \in X\}$$

is bounded in F for every $x \in E$. The importance of the classical Banach-Steinhaus theorem led Bourbaki to take it as a motivation for the following classical definition.

DEFINITION 4. E is a *barreled space* if, for every F, we have that each collection $X \subset L(E; F)$ is equicontinuous if (and always only if) X is pointwise bounded on E.

Remark 10. In Definition 4, use is made of an arbitrary F and of $L(E; F)$. It can be shown that it is enough to use only $F = \mathbb{C}$ in Definition 4. Moreover, it can be seen that the following definition, which involves only E, is equivalent to Definition 4; namely, E is a barreled space if each seminorm on E which is lower semicontinuous must be continuous.

Remark 11. By the classical Banach-Steinhaus theorem, the simplest criterion is the following: If E is a Banach space, or more generally, a Fréchet space, then it is barreled.

Continuous linear mappings and pointwise boundedness get along very well, in the sense that there is a fruitful interplay between the two concepts. However, the same is not true in the holomorphic case in the following sense: It may occur, by a classical example, that a sequence $f_m \in H(\mathbb{C}; \mathbb{C})$, for $m \in \mathbb{N}$, may be pointwise bounded on $\mathbb{C}$, that is,

$$\sup_{m \in \mathbb{N}} |f_m(x)| < +\infty$$

for every $x \in \mathbb{C}$, and yet not be bounded on some compact subset of $\mathbb{C}$, say

$$\sup_{\substack{m \in \mathbb{N} \\ |x| \leq 1}} |f_m(x)| = +\infty.$$

Since holomorphic mappings and pointwise boundedness do not get along very well, in the sense that we have just described, it is timely for our present considerations that Definition 4 is equivalent to the following Definition 5, which is not usually pointed out in the current expositions of topological vector spaces.

DEFINITION 5. E is a *barreled space* if, for every F, we have that each collection $X \subset L(E; F)$ is equicontinuous if (and always only if) X is bounded on all finite dimensional compact subsets of E.

Remark 12. As in Remark 10, we point out that it can be shown that it is enough to use only $F = \mathbb{C}$ in Definition 5.

Holomorphic mappings and boundedness on finite dimensional compact subsets get along very well, in the sense that there is a fruitful interplay between the two concepts. As a result, we shall pick up Definition 5 as a motivation for the following Definition 6.

DEFINITION 6. E is a *holomorphically barreled space* if, for every U and every F, we have that each collection $X \subset H(U; F)$ is equicontinuous if (and always only if) X is bounded on all finite dimensional compact subsets of U.

Remark 13. By comparing Definitions 5 and 6, it is clear that a holomorphically barreled space is also a barreled space.

PROPOSITION 4. E is a holomorphically barreled space if and only if, for every $U \subset E$, we have that each collection $X \subset H(U; \mathbb{C})$ is equicontinuous if X is bounded on all finite dimensional compact subsets of U.

Remark 14. Proposition 4 is to be compared with Remark 12.

PROPOSITION 5. A Baire space E is holomorphically barreled.

We recall that E is said to be a Baire space if every denumerable intersection of dense open subsets of E must be dense in E.

Remark 15. Proposition 5 is an improvement of the known fact that a Baire space E is barreled.

Proposition 5 contains as a particular case the following generalization of the classical Banach-Steinhaus theorem.

PROPOSITION 6 (Holomorphic Banach-Steinhaus Theorem). If E is a Fréchet space, in particular a Banach space, each collection $X \subset H(U; F)$ is equicontinuous if X is bounded on all finite dimensional compact subsets of U.

PROPOSITION 7. A Silva space E is holomorphically barreled.

The proof of Proposition 7 rests on the preceding Lemma 1.

Remark 16. Proposition 7 is a reminiscence of the known fact that any inductive limit of barreled spaces is barreled. In this connection, it is to be noticed that a denumerable inductive limit of holomorphically barreled spaces may fail to be holomorphically barreled.

The history of the concept of infrabarreled spaces is the following.
Let us consider the evaluation mapping

$$x \in E \longmapsto \tilde{x} \in L(L(E; F); F)$$

which is defined by

$$\tilde{x}(f) = f(x) \in F$$

for every $f \in L(E; F)$. In order that it be continuous, where

$$L(E; F) \quad \text{and} \quad L(L(E; F); F)$$

carry their strong topologies, it is necessary and sufficient that each collection $X \subset L(E; F)$ which is bounded on all bounded subsets of E be equicontinuous. The simple examples showing that such a necessary and sufficient condition is not always satisfied are due to the fact that E does not behave properly, in the sense that its topology is not large enough with respect to its collection of bounded subsets; moreover, there is a clear analogy between the above necessary and sufficient condition and Definition 5. This leads us to put our finger on E, and set the following classical definition.

DEFINITION 7. E is an *infrabarreled space* if, for every F, we have that each collection $X \subset L(E; F)$ is equicontinuous if (and always only if) X is bounded on all bounded subsets of E.

Remark 17. In Definition 7, use is made of an arbitrary F and of $L(E; F)$. It can be shown that is enough to use only $F = \mathbb{C}$ in Definition 7. By going back to the preceding motivation, E being infrabarreled amounts to the continuity of the evaluation mapping

$$x \in E \longmapsto \tilde{x} \in E''$$

of E into its strong double dual space. Moreover, it can be seen that the following definition, which involves only E, is equivalent to Definition 7; namely, E is an infrabarreled space if each seminorm on E which is lower semicontinuous and is bounded on every bounded subset of E must be continuous.

Since holomorphic mappings and bounded subsets do not get along very well, in the sense that we described in Sec. 2, it is fortunate for our present considerations that Definition 7 is equivalent to the following Definition 8, which is not usually pointed out in the current expositions of topological vector spaces.

DEFINITION 8. E is an *infrabarreled space* if, for every F, we have that each collection $X \subset L(E; F)$ is equicontinuous if (and always only if) X is bounded on all compact subsets of E.

Remark 18. As in Remark 17, we point out that it can be shown that it is enough to use only $F = \mathbb{C}$ in Definition 8. Moreover, it can be seen that the following definition, which involves only E, is equivalent to Definition 8; namely, E is an infrabarreled space if each seminorm on E which is lower semi-continuous and is bounded on every compact subset of E must be continuous.

Holomorphic mappings and compact subsets get along very well, in the sense that we described in Sec. 2. As a result, we shall pick up Definition 8 as a motivation of the following Definition 9.

DEFINITION 9. E is a *holomorphically infrabarreled space* if, for every U and every F, we have that each collection $X \subset H(U; F)$ is equicontinuous if X is bounded on all compact subsets of U.

Remark 19. By comparing Definitions 8 and 9, it is clear that a holomorphically infrabarreled space is also an infrabarreled space.

PROPERTIES OF LOCALLY CONVEX SPACES

PROPOSITION 8. E is a holomorphically infrabarreled space if and only if, for every $U \subset E$, we have that each collection $X \subset H(U; \mathbb{C})$ is equicontinuous if X is bounded on all compact subsets of U.

Remark 20. Proposition 8 is to be compared with Remarks 17 and 18.

PROPOSITION 9. A holomorphically barreled space E is holomorphically infrabarreled.

Remark 21. It is clear that a barreled space E is infrabarreled. Proposition 9 corresponds to this remark.

PROPOSITION 10. A holomorphically bornological space E is holomorphically infrabarreled.

Remark 22. It is known that a bornological space is infrabarreled. Proposition 10 corresponds to this remark.

PROPOSITION 11. E is a holomorphically bornological space if and only if E is a holomorphically infrabarreled space and, for every U, we have that each function $f: U \to \mathbb{C}$ belongs to $H(U; \mathbb{C})$ if (and always only if) f is algebraically holomorphic, and f is bounded on every compact subset of U.

Remark 23. Proposition 11 is a strengthening of Proposition 10, and it should be compared with Example 5.

Remark 24. It is known that E is a bornological space if and only if E is an infrabarreled space and each linear form f: E → $\mathbb{C}$ is continuous if (and always only if) f is bounded on every compact, or bounded, subset of E. Proposition 11 corresponds to this remark.

Remark 25. It is known that E is a barreled space if and only if E is an infrabarreled space and a certain Alaoglu-Bourbaki compactness property holds. The corresponding result in the present context was discussed by Matos [20], to read that E is a holomorphically barreled space if and only if E is a holomorphically infrabarreled space and E has a Montel compactness property. Such a result is a strengthening of Proposition 9, in a sense comparable to the case of Propositions 10 and 11; see Remark 23.

Remark 26. The analogy between barreled spaces and infrabarreled spaces is clear (compare Definitions 5 and 8; or the final parts of Remarks 10 and 17, 18). There is also an analogy between bornological spaces and infrabarreled spaces (compare the final parts of Remarks 1, 3 and 17, 18; or Definitions 2 and 8, which is more subtle); in view of such an analogy, infrabarreled spaces could also justifiedly be called infrabornological spaces. Such analogies in the linear case prevail in the holomorphic case as well; but we shall not elaborate on them.

Example 6. Consider the same space E as in Example 4. It is barreled as the Cartesian product of two barreled spaces. However, it is not holomorphically infrabarreled, hence it is not holomorphically barreled. In fact, introduce f_k: E → $\mathbb{C}$ defined by

$$f_k(x, y) = x_k y_k;$$

then $f_k \in H(E; \mathbb{C})$ for $k \in \mathbb{N}$. Since $f_k \to 0$ uniformly on every compact subset of E as $k \to \infty$, we see that

$$X = \{f_k; k \in \mathbb{N}\}$$

is bounded on every compact subset of E; but X is not equicontinuous. Thus we see that E is not holomorphically infrabarreled. Hence, a barreled space may fail to be holomorphically infrabarreled. We also see that the Cartesian product of two holomorphically barreled spaces may fail to be holomorphically infrabarreled. This example also shows that a denumerable direct sum, or a denumerable inductive limit, of holomorphically barreled spaces does not have to be holomorphically infrabarreled.

4. HOLOMORPHICALLY MACKEY SPACES

The history of the concept of Mackey spaces is the following.

Let be given a locally convex space E, whose topology we shall denote by T. We have its topological dual space E', the vector space of all continuous linear forms on E. Since E' is determined by T, we may ask if, conversely, E' determines T. The answer to this question is the following. Once a vector subspace E' of the algebraic dual space of a vector space E is given, there are two (necessarily unique) topologies $\sigma(E, E')$ and $\tau(E, E')$ on E such that $\sigma(E, E') \subset \tau(E, E')$, having the following property: a locally convex topology T on E determines E' as the topological dual space if and only if

$$\sigma(E, E') \subset T \subset \tau(E, E');$$

then $\sigma(E, E')$ is the weak topology determined on E by E', and $\tau(E, E')$ is the Mackey topology determined on E by E'. By the

Mackey-Arens theorem, $\tau(E, E')$ is the topology on E of uniform convergence on the subsets of E' that are convex and compact for the weak topology $\sigma(E', E)$ determined on E' by E. Finally, a locally convex space E is a Mackey space if its given topology coincides with $\tau(E, E')$, where E' is its topological dual space; in other words, the given topology on E is the largest locally convex topology on E giving rise to the same topological dual space E'. It is timely for our present considerations that the above outlook on Mackey spaces is equivalent to the following Definition 10, which is usually formulated as a proposition in the current expositions of topological vector spaces.

DEFINITION 10. E is a *Mackey space* if, for every F, we have that each mapping $f: E \to F$ belongs to $L(E; F)$ if (and always only if) f is linear, and f is weakly continuous, that is, $\psi \circ f$ is continuous for every $\psi \in F'$; in other notation,

$$L(E; F) = L(E; wF).$$

We recall that wF denotes F endowed with the weak topology $\sigma(F, F')$.

Remark 27. In Definition 10, use is made of an arbitrary F and of $L(E; F)$. Since E may fail to be a Mackey space, it is not enough to use only $F = \mathbb{C}$ in Definition 10.

Holomorphy being identical to weak holomorphy is a very desirable condition. As a result, we shall pick up Definition 10 as the motivation for the following Definition 11.

DEFINITION 11. E is a *holomorphically Mackey space* if, for every U and every F, we have that each mapping $f: U \to F$ belongs to H(U; F) if (and always only if) f is weakly holomorphic, that is, $\psi \circ f$ is holomorphic for every $\psi \in F'$; in other notation

$$H(U; F) = H(U; wF).$$

We point out that we have just used the sheaf H in place of the sheaf $\mathcal{H}$; see [22]. We can use $\mathcal{H}$ intead of H in Definition 11 without affecting it, provided we insist that F be complete; or even be complete in some of the known milder senses.

Remark 28. By comparing Definitions 10 and 11, and since continuous linear mappings are holomorphic, it is clear that a holomorphically Mackey space is a Mackey space.

PROPOSITION 12. A holomorphically infrabarreled space E is a holomorphically Mackey space.

Remark 29. It is known that an infrabarreled space is a Mackey space. Proposition 12 corresponds to this remark.

Remark 30. We can make comments analogous to Remark 25, with Mackey now in place of infrabarreled; but we shall not elaborate on them.

Remark 31. Proposition 11 remains true if we replace holomorphically infrabarreled by holomorphically Mackey. We can then make comments analogous to Remarks 23 and 24; but we shall not elaborate on them.

Example 7. Consider the same space E as in Examples 4 and 6. It can be seen that it is a holomorphically Mackey space.

5. APPENDIX

Given an infinite dimensional normed space E, up to recently it was an open question since Banach time to know whether, sequentially, the weak topology on E' defined by E and the norm topology on E' are always different. That this is indeed the case was proved in 1973, independently by Josefson [17, 18], from Sweden, and Nissenzveig [23], from Israel. We then have the following fundamental result.

THEOREM (Josefson-Nissenzveig). If E is an infinite dimensional normed space, there is a sequence f_m (m = 1, 2, ...) of continuous linear forms on E such that

$$f_m(x) \to 0$$

as $m \to \infty$ for any $x \in E$, and yet

$$\|f_m\| = 1$$

for all m = 1, 2, ...

This result has several applications in infinite dimensional holomorphy; one of them, which is of present interest, is the following

PROPOSITION. If E is an infinite dimensional complex normed space, there is an entire complex-valued function on E which is unbounded on some bounded subset of E.

The proof goes as follows (compare with Example 1). Choose the sequence (f_m) as in the above theorem. Define

$$f(x) = \sum_{m=1}^{\infty} [f_m(x)]^m$$

for every $x \in E$. Then $f \in H(E; \mathbb{C})$. If f were bounded on every bounded subset of E, the Cauchy inequality would imply that

$$\|f_m\| \to 0$$

as $m \to \infty$, which is not true; hence the desired result.

ACKNOWLEDGMENTS

The author gratefully acknowledges partial support from Universidade Federal do Rio de Janeiro (COPERTIDE), Financiadora de Estudos e Projetos (FINEP), Indústrias Klabin, and Centro Brasileiro de Pesquisas Físicas, Rio de Janeiro, GB, Brasil.

BIBLIOGRAPHY

1. Books on Locally Convex Spaces

1. N. Bourbaki, *Espaces Vectoriels Topologiques*, Hermann Éditeurs.
2. A. Grothendieck, *Topological Vector Spaces*, Gordon and Breach.
3. J. Horvath, *Topological Vector Spaces and Distributions*, Addison-Wesley.
4. G. Köthe, *Topological Vector Spaces*, Springer-Verlag.
5. H. H. Schaeffer, *Topological Vector Spaces*, Springer-Verlag.
6. F. Trèves, *Topological Vector Spaces, Distributions, and Kernels*, Academic Press.

2. Books on Infinite Dimensional Holomorphy

7. N. Bourbaki, *Variétés différentielles et analytiques, Fascicule de résultats*, 1-7 et 8-15, Hermann Editeurs,
8. G. Coeuré, Analytic functions and manifolds in infinite dimensional spaces, *Notas de Matemática*, Vol. 52, North-Holland.
9. L. Nachbin, Topology on spaces of holomorphic mappings, *Ergebnisse der Mathematik*, Bd. 47, Springer-Verlag.
10. Ph. Noverraz, Pseudo-convexité, convexité polynomiale, et domaines d'holomorphie en dimension infinie, *Notas de Matemática*, Vol. 48, North-Holland.
11. J.-P. Ramis, Sous-ensembles analytiques d'une variété banachique complexe, *Ergebnisse der Mathematik*, Bd. 53, Springer-Verlag.
12. Proceedings on Infinite Dimensional Holomorphy, 1973, *Lecture Notes in Mathematics*, Vol. 364, Springer-Verlag, Berlin, 1974.

3. Special References to Articles

13. J. A. Barroso, M. C. Matos, and L. Nachbin, On bounded sets of holomorphic mappings: see [12], pp. 123-134.
14. S. Dineen, Holomorphic functions on locally convex spaces, *Annales de l'Institut Fourier*, 23, 19-54 (I), 153-185 (II) (1973).
15. S. Dineen, Holomorphically significant properties of topological vector spaces, *Fonctions Analytiques de Plusieurs et Analyse Complexe, Colloque International du Centre National de la Recherche Scientifique, Paris, 1972*, Gauthier-Villars,
16. B. Josefson, A counterexample in the Levi problem: see [12], pp. 168-177.
17. B. Josefson, Weak sequential convergence in the dual of a Banach space does not imply norm convergence, *Bull. Am. Math. Soc.*, 81, 166-168 (1975).
18. B. Josefson, Weak sequential convergence in the dual of a Banach space does not imply norm convergence, *Arkiv für Matematik* 13, 79-89 (1975).
19. P. Lelong, Théorème de Banach-Steinhaus pour les polynômes: applications analytiques entières d'espaces vectoriels, Séminaire Pierre Lelong 1970, *Lecture Notes in Mathematics*, Vol. 205, Springer-Verlag, Berlin, 1971, 87-112.
20. M. C. Matos, On locally convex spaces with the Montel property, this volume, pp. 245-250.
21. L. Nachbin, Recent developments in infinite dimensional holomorphy, *Bull. Am. Math. Soc.*, 79, 625-640 (1973).
22. L. Nachbin, A glimpse at infinite dimensional holomorphy: see [12], pp. 69-79.
23. A. Nissenzveig, W^* sequential convergence, *Israel J. Math.* (to appear).

REMARKS ON NONLINEAR EVOLUTION EQUATIONS IN NONCYLINDRICAL DOMAINS[*]

B. P. Neves and L. A. Medeiros

Instituto de Matemática
Universidade Federal do Rio de Janeiro
Rio de Janeiro, Brazil

In this paper we generalize for noncylindrical domains a regularity result proved by Lions [1] Theorem 1.3 in the noncylindrical case. We shall use the penalty method as in Lions [2], plus Galerkin approximations, choosing for base the eigenfunctions of the Laplace operator. Before we formulate our result (see Theorem 1), we give some results about the function spaces which we shall use. We did this research when we were at the Department of Mathematics at Tulane University, during the Winter Quarter of 1974.

[*]This research was supported by FINEP, CEPG-UFRJ, and by C. B. P. F. (second author).

1. FUNCTION SPACES

Let Ω be an open set of the Euclidian space $\mathbb{R}^n$. All functions we consider are real and all derivatives are in the sense of the theory of distributions. By dx we represent the Lebesgue measure on $\mathbb{R}^n$, and by $L^p(\Omega)$, $1 \leq p \leq +\infty$, the Banach space of functions for which the power p is Lebesgue integrable in Ω. When $p = 2$ we have a Hilbert space with the inner product given by $(u, v) = \int_\Omega u(x)v(x)\,dx$. For each natural number m we define the vector space $H^m(\Omega)$ of all functions u in $L^2(\Omega)$ such that all derivatives of order less than or equal to m belong to $L^2(\Omega)$. We know that $H^m(\Omega)$ is a Hilbert space with the inner product given by

$$(u, v)_m = \sum_{|\alpha| \leq m} (D^\alpha u, D^\alpha v).$$

The Hilbert space $H^m(\Omega)$ is called a Sobolev space. By $H_0^m(\Omega)$ we represent the closure of $C_0^\infty(\Omega)$ in $H^m(\Omega)$ with the norm induced by the inner product $(\ ,\ )_m$. We note by $C_0^\infty(\Omega)$ the space of C^∞ functions with compact support in Ω. It is important to observe that when Ω has a smooth boundary, it follows that if u belongs to $H_0^m(\Omega)$, then u and all its derivatives up to the order m are zero on the boundary of Ω. For all those results on Sobolev spaces, see Lions [3]. If X is a Banach space, we represent by $L^p(0, T; X)$, $1 \leq p \leq +\infty$, the space of strongly measurable functions $t \to f(t)$ from $(0, T)$ into X such that $\|f(t)\|_X$ belongs to $L^p(0, T)$. This is a Banach space with the norm

$$\|f\|^p_{L^p(0,T;X)} = \int_0^T \|f(t)\|^p_X\,dt$$

if $p \neq \infty$ and with the usual change when $p = \infty$.

NONLINEAR EVOLUTION EQUATIONS

2. REGULARITY IN NONCYLINDRICAL DOMAINS

Let us consider the strip

$$B = \{(x, t); x \in \mathbb{R}^n, 0 \leq t \leq T\}$$

in $\mathbb{R}^n$ and let Q be a bounded open set in B. Represent by Ω_s the intersection of Q with the hyperplane $P_s = \{(x, t) \in B; t = s\}$. By Ω_0 and Ω_T we represent the intersections of Q with P_0 and P_T respectively. We represent by Γ_s the boundary of Ω_s and by $\Sigma = \bigcup_{0 \leq t \leq T} \Gamma_t$ the lateral boundary of Q. Hypothesis 2 below says that Σ is smooth.

HYPOTHESIS 1. Ω_t grows with t in (0, T), that is, if Ω_t^* is the projection of Ω_t on the hyperplane P_0, then $\Omega_t^* \subset \Omega_s^*$ if $t \leq s$.

HYPOTHESIS 2. Let $\Omega \times (0, T)$ be a cylinder that contains Q. If for almost all $t \in (0, T)$, $u(t) \in H^1(\Omega)$ and $u = 0$ almost every where in $\Omega - \Omega_t$, then the restriction of u to Ω_t belongs to $H_0^1(\Omega_t)$ for almost all t in (0, T).

With the above notations and assumptions, we can formulate our result in the following theorem:

THEOREM 1. Suppose the hypotheses 1 and 2. Let f, u_0, u_1 be given functions such that

$$f \in H_0^1(Q), \tag{1}$$

$$u_0 \in H_0^2(\Omega_0), \tag{2}$$

$$u_1 \in H_0^1(\Omega_0). \tag{3}$$

Then, if $\rho \leq 2/(n - 2)$, $n \geq 3$, there exists a function u defined on Q satisfying the following conditions:

$$u \in L^{\infty}(0, T; H_0^1(\Omega_t) \cap H^2(\Omega_t)); \qquad (4)$$

$$u_t \in L^{\infty}(0, T; H_0^1(\Omega_t)); \qquad (5)$$

$$u_{tt} \in L^{\infty}(0, T; L^2(\Omega_t)); \qquad (6)$$

$$u_{tt} - \Delta u + |u|^{\rho} u = f \qquad (7)$$

in the sense of the theory of distributions in Q;

$$u(x, 0) = u_0(x), \; u_t(x, 0) = u_1(x) \quad \text{on } \Omega_0. \qquad (8)$$

The proof of Theorem 1 will be given by the penalty method. Represent by Ω a bounded open set $\mathbb{R}_x^n$, with smooth boundary such that Q is contained in $\Omega \times [0, T]$. Let M be the function equal to zero on Q and 1 in $\Omega \times [0, T] - Q$. Let F be the extension of f to the cylinder $\Omega \times [0, T]$, zero outside Q. Represent by U_0, U_1 the extensions of u_0, u_1 defined by zero outside Ω_0. With these extensions we change the problem from the open set Q to the cylinder $\Omega \times [0, T]$. For each $\varepsilon > 0$, we consider the equation

$$U^{\varepsilon\prime\prime} - \Delta U^{\varepsilon} + |U^{\varepsilon}|^{\rho} U^{\varepsilon} + \frac{1}{\varepsilon} M U^{\varepsilon\prime} = F$$

on $\Omega \times [0, T]$. We prove that there exists a solution U for each $\varepsilon > 0$, with certain regularity, such that $U^{\varepsilon} \to U$ as $\varepsilon \to 0$ and that the restriction of U to Q is the solution claimed in Theorem 1. This argument is based in the following lemma.

LEMMA 1. For each $\varepsilon > 0$ there exists a function $U^{\varepsilon} : \Omega \times [0, T] \to \mathbb{R}^n$ satisfying the following conditions:

$$U^{\varepsilon} \in L^{\infty}(0, T; H_0^1(\Omega) \cap H^2(\Omega)) \qquad (9)$$

and belongs to a bounded set of this space when $\varepsilon \to 0$;

$$U^{\varepsilon'} \in L^{\infty}(0, T; H_0^1(\Omega)) \tag{10}$$

and belongs to a bounded set of this space when $\varepsilon \to 0$;

$$U^{\varepsilon''} \in L^{\infty}(0, T; L^2(\Omega)), \tag{11}$$

$$U^{\varepsilon''} - \Delta U^{\varepsilon} + |U^{\varepsilon}|^{\rho} U^{\varepsilon} + \frac{1}{\varepsilon} M U^{\varepsilon'} = F \tag{12}$$

in the sense of the theory of distributions on $\Omega \times [0, T]$;

$$U^{\varepsilon}(x, 0) = U_0(x), \quad U^{\varepsilon'}(x, 0) = U_1(x) \quad \text{on } \Omega. \tag{13}$$

Proof of Lemma 1. Let $(\omega_j)_{j \in \mathbb{N}}$ be the eigenfunctions for $-\Delta$ on Ω, that is,

$$-\Delta \omega_j = \lambda_j \omega_j, \quad \omega_j = 0 \quad \text{on } \partial\Omega. \tag{14}$$

We know that $\omega_j \in H_0^1(\Omega) \cap H^2(\Omega)$.

Choose U_{0m}, U_{1m} in the subspace $[\omega_1, \omega_2, \ldots, \omega_n]$ such that

$$U_{0m} \to U_0 \quad \text{strongly in} \quad H_0^1(\Omega) \cap H^2(\Omega), \tag{15}$$

$$U_{1m} \to U_1 \quad \text{strongly in} \quad H_0^1(\Omega). \tag{16}$$

Let us consider the functions

$$U_m^{\varepsilon}(x, t) = \sum_{j=1}^{m} g_{jm}(t) \omega_j(x)$$

with $g_{jm} \in C^2[0, T]$, determined by the conditions

$$(U_m^{\varepsilon''}(t), \omega_j) + a(U^{\varepsilon}(t), \omega_j)$$
$$+ (|U_m^{\varepsilon}|^{\rho} U_m^{\varepsilon}, \omega_j) + \frac{1}{\varepsilon} (M U_m^{\varepsilon}(t), \omega_j)$$
$$= (F(t), \omega_j) \tag{17}$$

for $1 \leq j \leq m$, where

$$a(u, v) = \sum_{i=1}^{n} \int_{\Omega} u_{x_i} v_{x_i} \, dx,$$

with the initial conditions

$$U_m^\varepsilon(0) = U_{0m}, \qquad (18)$$

$$U_m^{\varepsilon'}(0) = U_{1m}. \qquad (19)$$

From the hypothesis on ω_j, it follows that the system (17) of ordinary differential equations, with initial data, has a solution on an interval $[0, t_m]$. Therefore, the a priori estimates to be proved in the following show that the solution there exists on the interval $[0, T]$ with T independent of m.

A Priori Estimates-I. Multiply (17) by $g'_{jm}(t)$ and add in j from 1 to m. We obtain

$$\frac{1}{2} \frac{d}{dt} \{ |U_m^{\varepsilon'}(t)|^2 + \|U_m(t)\|^2 \}$$

$$+ \frac{1}{p} \frac{d}{dt} \|U_m(t)\|_{L^p(\Omega)}^p + \frac{1}{\varepsilon}(MU_m^{\varepsilon'}(t), U^{\varepsilon'}(t))$$

$$= (F(t), U_m^{\varepsilon'}(t)). \qquad (20)$$

Integrating from 0 to t and using (18) and (19), we obtain

$$\frac{1}{2} \{ |U_m^{\varepsilon'}(t)|^2 + \|U_m^\varepsilon(t)\|^2 \} + \frac{1}{p} \|U_m^\varepsilon(t)\|_{L^p(\Omega)}^p$$

$$+ \frac{1}{\varepsilon} \int_0^t (MU_m^{\varepsilon'}, U_m^{\varepsilon'}) \, ds \leq k + \frac{1}{2} \int_0^t |U_m^{\varepsilon'}(s)|^2 \, ds. \qquad (21)$$

From Gronwal's lemma we have

$$\|U_m^\varepsilon(t)\| < C, \qquad (22)$$

$$|U_m^{\varepsilon'}(t)| < C, \qquad (23)$$

$$\|U_m^{\varepsilon}(t)\|_{L^p(\Omega)} < C, \qquad (24)$$

all these inequalities are independent of ε and m:

$$\int_{\Omega \times [0,T]} M(U^{\varepsilon'}(x, t))^2 \, dx \, dt < C\varepsilon. \qquad (25)$$

A Priori Estimates-II. Put $\Delta \omega_j$ instead of $-\lambda_j^{-1} \omega_j$ in (17), multiply by $g'_{jm}(t)$, and add in j from 1 to m. We obtain

$$a(U_m^{\varepsilon''}(t), U_m^{\varepsilon'}(t)) + (\Delta U_m^{\varepsilon}(t), \Delta U_m^{\varepsilon'}(t))$$

$$+ (\rho + 1) \sum_{i=1}^{n} \left(|U_m^{\varepsilon}(t)|^{\rho} \frac{\partial U_m^{\varepsilon}(t)}{\partial x_i}, \frac{\partial U_m^{\varepsilon'}(t)}{\partial x_i} \right)$$

$$+ \frac{1}{\varepsilon} (M U_m^{\varepsilon'}(t), -\Delta U_m^{\varepsilon'}(t)) = a(F(t), U_m^{\varepsilon'}(t)). \qquad (26)$$

From the Hölder inequality we obtain

$$\left| \left(|U_m(t)|^{\rho} \frac{\partial U_m^{\varepsilon}(t)}{\partial x_i}, \frac{\partial U_m^{\varepsilon'}(t)}{\partial t} \right) \right|$$

$$\leq \| |U_m(t)|^{\rho} \|_{L^n(\Omega)} \left\| \frac{\partial U_m^{\varepsilon}(t)}{\partial x_i} \right\|_{L^q(\Omega)} \left\| \frac{\partial U_m^{\varepsilon'}(t)}{\partial x_i} \right\|_{L^2(\Omega)},$$

where q is given by $(1/q) + (1/n) + (1/2) = 1$, as in the embedding Sobolev's theorem. We have $\rho n \leq q$ because $\rho \leq 2/(n-2)$, $n \geq 3$. Then it follows that

$$\| |U_m^{\varepsilon}(t)|^{\rho} \|_{L^n(\Omega)} \leq C_1 \|U_m^{\varepsilon}(t)\|^{\rho} \leq \text{constant}$$

by (22);

$$\left\| \frac{\partial U_m^{\varepsilon}(t)}{\partial x_i} \right\|_{L^q(\Omega)} \leq C_2 \left\| \frac{\partial U_m^{\varepsilon}(t)}{\partial x_i} \right\|_{H_0^1(\Omega)} \leq C_2 \|U_m^{\varepsilon}(t)\|_{H^2(\Omega)}$$

$$\leq C_3 |\Delta U_m^{\varepsilon}(t)|_{L^2(\Omega)}$$

and

$$\left|\frac{\partial U_m^{\varepsilon'}}{\partial x_i}\right|_{L^2(\Omega)} \leq \|U_m^{\varepsilon'}(t)\|_{H_0^1(\Omega)}.$$

Therefore,

$$\sum_{i=1}^n \left|\left(|U_m^{\varepsilon}(t)|^\rho \frac{\partial U_m^{\varepsilon}(t)}{\partial x_i}, \frac{\partial U_m^{\varepsilon'}(t)}{\partial x_i}\right)\right| \leq nC\|U_m^{\varepsilon'}(t)\||\Delta U_m^{\varepsilon}(t)|.$$

Substituting in (26) we get

$$\frac{d}{dt}\{\|U_m^{\varepsilon'}(t)\|^2 + |\Delta U_m(t)|^2\} + \frac{1}{\varepsilon}(M\,U_m^{\varepsilon'}(t), -\Delta U_m^{\varepsilon'}(t))$$

$$\leq \|F(t)\|\|U_m^{\varepsilon'}(t)\| + nC\|U_m^{\varepsilon'}(t)\||\Delta U_m^{\varepsilon}(t)|. \qquad (27)$$

Integrating from 0 to t, we have

$$\frac{1}{2}\{\|U_m^{\varepsilon}(t)\|^2 + |\Delta U_m(t)|^2\} + \frac{1}{\varepsilon}\int_0^t (M\,U_m^{\varepsilon}(s), -\Delta U_m^{\varepsilon}(s))ds$$

$$\leq \frac{1}{2}\int_0^t \|F(s)\|^2\,ds + \frac{1}{2}\int_0^t \|U_m^{\varepsilon'}(s)\|^2\,ds$$

$$+ \frac{nC}{2}\int_0^t \|U_m^{\varepsilon'}(t)\|^2\,ds + \frac{nC}{2}\int_0^t |\Delta U_m^{\varepsilon}(s)|^2\,ds$$

$$+ \frac{1}{2}\{\|U_{1m}\|^2 + |\Delta U_{0m}|^2\}. \qquad (28)$$

Since

$$(M\,U_m^{\varepsilon'}(s), -\Delta U_m^{\varepsilon'}(s)) = \left(M\,U_m^{\varepsilon'}(s), \sum_{j=1}^m g_{jm}'(s)\,\lambda_j\omega_j\right)$$

$$\geq \inf\{\lambda_j\}(M\,U_m^{\varepsilon'}(s), U_m^{\varepsilon'}(s)) \geq 0,$$

we can write (28) in the following form

$$\|U_m^{\varepsilon'}(t)\|^2 + |\Delta U_m^{\varepsilon'}(t)|^2$$

$$\leq C_1 + C_2 \int_0^t \{\|U_m^{\varepsilon'}(s)\|^2 + |\Delta U_m^{\varepsilon}(s)|^2\}\,ds, \qquad (29)$$

where C_1 and C_2 are independent of m and ε. It follows that

$$\|U_m^{\varepsilon'}(t)\| \leq C, \quad \text{independent of } m \text{ and } \varepsilon; \qquad (30)$$

$$|U_m^{\varepsilon}(t)| \leq C, \quad \text{independent of } m \text{ and } \varepsilon. \qquad (31)$$

From (22)-(24), (30), and (31), we can say that

$$U_m^{\varepsilon} \text{ belongs to a bounded set } B_1 \text{ of}$$
$$L^{\infty}(0, T; H_0^1(\Omega) \cap H^2(\Omega)); \qquad (32)$$

$$U_m^{\varepsilon'} \text{ belongs to a bounded set } B_2 \text{ of}$$
$$L^{\infty}(0, T; H_0^1(\Omega)); \qquad (33)$$

$$|U_m^{\varepsilon}|^{\rho} U_m^{\varepsilon} \text{ belongs to a bounded set } B_3 \text{ of}$$
$$L^{\infty}(0, T; L^{P'}(\Omega)), \quad P'(\rho + 1) = \rho + 2. \qquad (34)$$

Observe that B_1, B_2, B_3 are independent of ε and m. Therefore, from (32)-(34) we can extract a subsequence U_υ^{ε} of U_m^{ε} such that:

$$U_\upsilon^{\varepsilon} \to U^{\varepsilon} \quad \text{weak star in} \quad L^{\infty}(0, T; H_0^1(\Omega) \cap H^2(\Omega)); \qquad (35)$$

$$U_\upsilon^{\varepsilon'} \to U^{\varepsilon'} \quad \text{weak star in} \quad L^{\infty}(0, T; H_0^1(\Omega)); \qquad (36)$$

$$|U_\upsilon^{\varepsilon}|^{\rho} U_\upsilon^{\varepsilon} \to \chi \quad \text{weak star in} \quad L^{\infty}(0, T; L^{P'}(\Omega)). \qquad (37)$$

From (35) and (36), the Rellich compactness theorem, and Lemma 1.3 in Lions [1], we conclude that

$$\chi = |U^{\varepsilon}|^{\rho} U^{\varepsilon}. \qquad (38)$$

Now, take $m = \upsilon$ in (17) and let $\upsilon \to \infty$ for $j \leq \upsilon$ fixed; we get

$$(U^{\varepsilon''}, \omega_j) + a(U^\varepsilon, \omega_j) + (|U^\varepsilon|^\rho U^\varepsilon, \omega_j) + \frac{1}{\varepsilon}(MU^{\varepsilon'}, \omega_j)$$

$$= (F(t), \omega_j). \tag{39}$$

Then

$$U^{\varepsilon''} - \Delta U^\varepsilon + |U^\varepsilon|^\rho U^\varepsilon + \frac{1}{\varepsilon} MU^\varepsilon = F \tag{40}$$

in the sense of $\mathcal{D}'(\Omega \times [0, T])$.

It follows from (32) and (33) that:

U belongs to a bounded set of

$$L^\infty(0, T; H_0^1(\Omega) \cap H^2(\Omega)) \quad \text{independent of } \varepsilon; \tag{41}$$

$U^{\varepsilon'}$ belongs to a bounded set of

$$L^\infty(0, T; H_0^1(\Omega)) \quad \text{independent of } \varepsilon. \tag{42}$$

From (40)-(42) and the hypothesis on F, it follows that

$$U^{\varepsilon''} \quad \text{belongs to} \quad L^\infty(0, T; L^2(\Omega)). \tag{43}$$

By standard argument, we can show that U^ε satisfies the initial data, which completes the proof of Lemma 1.

Proof of Theorem 1. As we have said above, to prove Theorem 1 it is sufficient to show that $U^\varepsilon \to U$ as $\varepsilon \to 0$ and that the restriction of U to Q, which we represent by u, satisfies all the conditions expressed in Theorem 1.

In fact, from (41) and (42) we can extract a sequence U_ν of U^ε such that

$$U_\nu \to U \quad \text{weak star in} \quad L^\infty(0, T; H_0^1(\Omega) \cap H^2(\Omega)); \tag{44}$$

$$U_\nu' \to U' \quad \text{weak star in} \quad L^\infty(0, T; H_0^1(\Omega)); \tag{45}$$

$$|U_\upsilon|^\rho U \to \chi \quad \text{weakly in} \quad L^\infty(0, T; L^2(\Omega)). \tag{46}$$

As in the proof of Lemma 1, if we use the Rellich compactness theorem and Lemma 1.3 of Lions [1], we obtain

$$\chi = |U|^\rho U.$$

Let u (respectively u) be the restriction of U^ε (of U) to Q. Let us consider the penalized equation (12) and take its restriction to Q. Since M = 0 on Q, we obtain

$$u_\varepsilon'' - \Delta u_\varepsilon + |u_\varepsilon|^\rho u_\varepsilon = f \quad \text{on } Q. \tag{47}$$

Take $\varepsilon = \upsilon$ and let $\upsilon \to \infty$ in the weak sense on Q; we obtain

$$u'' - \Delta u + |u|^\rho u = f \quad \text{on } \mathcal{D}'(Q). \tag{48}$$

We need to show that $u \in L^\infty(0, T; H_0^1(\Omega_t) \cap H^2(\Omega_t))$ and $u' \in L^\infty(0, T; H_0^1(\Omega_t))$. In fact, from (25) we have

$$\frac{1}{\sqrt{\upsilon}} MU_\upsilon' \to \Psi \quad \text{weakly in} \quad L^2(\Omega \times [0, T])$$

or

$$MU_\upsilon' \to 0 \quad \text{weakly in the same space.}$$

But $MU_\upsilon' \to MU'$ weak star in $L^\infty(0, T; L^2(\Omega))$, that is,

$$MU' = 0 \quad \text{a.e. in } \Omega \times [0, T]. \tag{49}$$

Since M = 0 on Q, we obtain $U' = 0$ a.e. in $\Omega \times (0, T) - Q$, and because Q is increasing and from Hypotheses 1 and 2 it follows that

$$u' \in L^\infty(0, T; H_0^1(\Omega_t)). \tag{50}$$

Still by $U' = 0$ a.e. in $\Omega \times (0, T)$ we obtain

$$U(x, t) = \omega(x) \quad \text{a.e. in } \Omega \times (0, T) - Q,$$

that is,

$$U(x, 0) = \omega(x) \quad \text{a.e. in} \quad \Omega - \Omega_0. \tag{51}$$

But

$$U(x, 0) = \lim_{\upsilon \to \infty} U_\upsilon(x, 0) = U(x) = 0$$

in $\Omega - \Omega_0$; therefore $\omega(x) = 0$ a.e. in $\Omega - \Omega_0$.

From this conclusion and Hypothesis 1, it follows that

$$U(x, t) = 0 \quad \text{a.e. in} \quad \Omega \times (0, T) - Q, \tag{52}$$

and from Hypothesis 2 we have

$$u \in L^\infty(0, T; H_0^1(\Omega_t) \cap H^2(\Omega_t)). \tag{53}$$

By (48), (50), and (53) we obtain

$$u'' \in L^\infty(0, T; L^2(\Omega_t)). \tag{54}$$

To check the initial data is standard argument. This completes the proof of the theorem.

REFERENCES

1. J. L. Lions, *Quelques Méthodes de Resolution des Problèmes aux Limites non Linéaires*, Dunod, Gauthier Villars, 1969.
2. J. L. Lions, Une remarque sur les problèmes d'evolution non lineaire dans des domaines non cylindriques, *Rev. Roumaine Math. Pure Appl.*, 9 (1964), 11-18.
3. J. L. Lions, *Problèmes aux Limites dans les Équations aux Derivées Partielles*, Les Press de l'Université de Montreal, 1965.

SOLUTIONS OF A NONLINEAR ABSTRACT CAUCHY-KOVALEWSKY SYSTEM AS A LOCAL BANACH ANALYTIC MANIFOLD

Domingos Pisanelli

Departamento de Matemática
Universidade de São Paulo
São Paulo, Brazil

This work is motivated by the theory of infinite Lie groups when we attempt to parametrize a subgroup given by a partial differential equation.

Theorem 1 shows that our hypotheses are equivalent to those of Treves in the generalization of Yamanaka-Ovcyannicov theorem, and are easier. This we show in Example 3 of the Cauchy-Kovalewsky partial differential equation, where we avoid the quasilinearization process of other authors.

Theorem 2 shows that solutions associated to the initial conditions (parameters) of the equation $dx/dt = f(t, x)$, $x(0) = \varphi$, when φ belongs to a ball of the space X_1 of the Banach scale $(X_s)_{0 < s \leq 1}$, can be considered a local Banach analytic manifold.

By definition, a *Banach scale* is a vector space over $\mathbb{C}$, which is the union of a family $(X_s)_{0 < s \leq 1}$ of Banach spaces such that

$$X_{s'} \supset X_s ; \ \| \ \|_{s'} \leq \| \ \|_s \Longleftarrow 0 < s' < s \leq 1.$$

Let H be the vector space over $\mathbb{C}$ of formal series $\sum_{n \geq 0} x_n T^n$, where $x_n \in \bigcap_{0 < s \leq 1} X_s$ and $x_0 \in X_1$.

Let Δ be a positive real number and

$$H_\Delta = \{x \in H \mid \sup_{0 < s \leq 1} \sup_{|t| < \Delta(1-s)} \|x(t)\|_s < +\infty\}.$$

By definition the series $x(t) = \sum_{n \geq 0} x_n t^n$ converges in the ball $|t| < \Delta(1 - s)$ with values in X_s ($\forall\ 0 < s < 1$). We have then a holomorphic function in $|t| < \Delta(1 - s)$ with values in X_s. When $s = 1$ we have, by definition,

$$\sup_{|t| < \Delta(1-1)} \|x(t)\|_1 = \|x_0\|_1.$$

Let

$$\|x\| = \sup_{0 < s \leq 1} \sup_{|t| < \Delta(1-s)} \|x(t)\|_s, \quad x \in H_\Delta.$$

We have $\|x^1 + x^2\| \leq \|x^1\| + \|x^2\|$, $\|\alpha x\| = |\alpha| \|x\|$ whenever, x^1, x^2, $x \in H_\Delta$, $\alpha \in \mathbb{C}$, i.e., H_Δ is a vector space over $\mathbb{C}$ and $\|x\|$ is a seminorm.

But $\|x\|$ is also a norm:

$0 = \|x\| \implies x(t) = \sum_{n \geq 0} x_n t^n = 0 \quad \forall\ |t| < \Delta(1 - s),\ 0 < s < 1.$

We then have $x_n = 0\ (n \geq 0)$.

H *is a Banach space:*

Let $(x^n)_{n \geq 1}$ be a Cauchy sequence in H_Δ. We have

$$\|x^n(t) - x^m(t)\|_s < \varepsilon \impliedby n, \tag{1}$$

$m \geq n_0(\varepsilon),\ \varepsilon > 0,\ |t| < \Delta(1 - s),\ 0 < s < 1;$

$$\|x_0^n - x_0^m\|_1 < \varepsilon \impliedby n,\ m \geq n_0(\varepsilon),\ \varepsilon > 0. \tag{2}$$

Inequality (1) gives us that x^n converges uniformly on $|t| < \Delta(1 - s)$ to x_s holomorphic with values in X_s.

Coefficients of x_s do not depend on s: if $s' < s$, x^n converges uniformly to $x_{s'}$, on $|t| < \Delta(1 - s')$, a *fortiori* on $|t| < \Delta(1 - s)$, with values in $X_{s'}$. But $X_s \hookrightarrow X_{s'}$, so that $x_s(t) = x_{s'}(t)$ when $|t| < \Delta(1 - s')$. Let x be the formal series whose coefficients are those of x_s. We have in X_s ($0 < s < 1$):

$$x_0 = \lim_{n \to \infty} x^n(0) = \lim_{n \to \infty} x_0^n.$$

Inequality (2) gives us $\lim_{n \to \infty} x_0^n \in X_1 \hookrightarrow X_s$, so that $x_0 \in X_1$.

We have from (1) and (2)

$$\|x^n(t) - x(t)\|_s \leq \varepsilon \iff n \geq n_0(\varepsilon),$$

$$\varepsilon > 0, \quad |t| < \Delta(1 - s), \quad 0 < s < 1;$$

$$\|x_0^n - x_0\|_1 \leq \varepsilon \iff n \geq n_0(\varepsilon), \quad \varepsilon > 0,$$

i.e., $x \in H_\Delta$ and $\lim_{n \to \infty} x^n = x$ in H_Δ.

Example 1. Let $z \in \mathbb{C}^p$, $|z| = \sup_{1 \leq j \leq p} |z_j|$, $\tau \in \mathbb{C}^n$, $|\tau| = \sup_{1 \leq j \leq n} |\tau_j|$, $B(0, s) = \{\tau \in \mathbb{C}^n \mid |\tau| < s\}$, and $X_s = H^1(B(0, s), \mathbb{C}^p)$ the Banach space of functions of n complex variables $\tau \in \mathbb{C}^n$, with values in $\mathbb{C}^p$, holomorphic and bounded, with partial derivatives with respect to τ_j ($1 \leq j \leq n$) bounded in the ball $B(0, s)$, with the norm

$$\|x\|_s = \sup_{|\tau|<s} \max_{1 \leq j \leq n} \left(|x(\tau)|, \left|\frac{\partial x}{\partial \tau_j}(\tau)\right| \right), \quad 0 < s < 1.$$

The space H_Δ constructed from $(X_s)_{0<s\leq 1}$ is isomorphic to the Banach space $H'(B_\Delta, \mathbb{C}^p)$ of functions x of n + 1 complex variables, $(t, \tau_1, \ldots, \tau_n)$, holomorphic, and bounded, with partial derivatives with respect to τ_j $(1 \leq j \leq n)$ bounded in B: $(|t|/\Delta) + |\tau| < 1$, endowed with the norm of the supremum of the function $|x|$ and the derivatives $|\partial x/\partial \tau_j|$ $(1 \leq j \leq n)$.

Let $x \in H_\Delta$. We can associate to it the holomorphic function of n + 1 complex variables $x(t, \tau) = \sum_{m \geq 0} x_m(\tau) t^m$, defined when $|t| < \Delta(1 - s)$, $|\tau| < s$, $\forall\ 0 < s < 1$, i.e., in $(|t|/\Delta) + |\tau| < 1$. We have $|x(t, \tau)| \leq \|x\|$ and $|(\partial/\partial \tau_j) x(t, \tau)| \leq \|x\|$ in B_Δ.

Conversely, let $(t, \tau) \in B_\Delta \to x(t, \tau)$ be holomorphic, bounded with partial derivatives with respect to τ_j $(1 \leq j \leq n)$ bounded by $M > 0$. We have that, for fixed $|\tau| < 1$:

$$x(t, \tau) = \sum_{m \geq 0} \frac{1}{m!} \frac{\partial^m}{\partial t^m} x(0, \tau), \quad |t| < \Delta(1 - |\tau|).$$

$$x_m(\tau) = \frac{1}{m!} \frac{\partial^m}{\partial t^m} x(0, \tau), \quad (m \geq 1),$$

is holomorphic in $|\tau| < 1$, bounded in $|\tau| < s$ with its partial derivatives with respect to τ_j $(1 \leq j \leq n)$ $(0 < s < 1)$ $x_0(\tau) = x(0, \tau)$ is holomorphic and bounded in $|\tau| < 1$ with its derivatives with respect to τ_j $(1 \leq j \leq n)$. We then have $x_m \in \bigcap_{0<s\leq 1} X_s$ and $x_0 \in X_1$.

The formal series $x = \sum_{m \geq 0} x_m T^m$ converges in $|t| < \Delta(1 - s)$ with values in X_s:

$$|x_m(\tau)| \leq \frac{1}{m!} \left| \frac{\partial^m}{\partial t^m} x(0, \tau) \right| \leq \frac{M}{[\Delta(1 - |\tau|)]^m} \leq \frac{M}{[\Delta(1 - s)]^m},$$

$$\left| \frac{\partial x_m}{\partial \tau_j}(\tau) \right| \leq \frac{1}{m!} \left| \frac{\partial^{m+1}}{\partial t^m \partial \tau_j} x(0, \tau) \right| \leq \frac{M}{[\Delta(1 - |\tau|)]^m} \leq \frac{M}{[\Delta(1 - s)]^m}$$

when $|\tau| < s$, i.e. $\|x_m\|_s \leq M/[\Delta(1-s)]^m$ ($\forall\ m \geq 0$) when $0 < s < 1$, which proves that $x(t) \in X_s$ when $|t| < \Delta(1-s)$, $0 < s < 1$.

But $\sup_{0<s\leq 1} \sup_{|t|<\Delta(1-s)} \|x(t)\|_s \leq M$, i.e., $x \in H_\Delta$.

The mapping $x \in H_\Delta \to ((t, \tau) \in B_\Delta \to x(t, \tau)) \in H^1(B_\Delta, \mathbb{C}^p)$ is an isometry for

$$\|x\| = \sup_{0<s\leq 1} \sup_{|t|<\Delta(1-s)} \|x(t)\|_s$$

$$= \sup_{0<s\leq 1} \sup_{|t|<\Delta(1-s)} \sup_{|\tau|<s} \max_{1\leq j\leq n} \left(|x(t)(\tau)|, \left|\frac{\partial}{\partial \tau_j} x(t)(\tau)\right|\right)$$

$$= \sup_{B_\Delta} \max_{1\leq j\leq n} \left(|x(t)(\tau)|, \left|\frac{\partial}{\partial \tau_j} x(t)(\tau)\right|\right).$$

Example 2. Let $X = \bigcup_{0<s\leq 1} X_s$ be a Banach scale and the system

$$\frac{dx}{dt} = f(t, x), \quad x(0) = x_0 \in X_1, \tag{3}$$

where:

(I) $$f(t, x) = \sum_{\substack{p\geq 0 \\ q\geq 0}} \hat{F}_{p,q}(x - x_0) t^q$$

and

$$\hat{F}_{p,q}(h) = F_{p,q}(h, \ldots, h).$$

(II) $F_{p,q}(h_1, \ldots, h_p)$ is p-linear in X_s^p with values in $X_{s'}$, ($0 < s' < s \leq 1$), $p > 0$, $q \geq 0$. $\hat{F}_{0,q} \in X$ when $q \geq 0$.

(III) $$\|F_{p,q}(h_1, \ldots, h_p)\|_{s'} \leq \frac{c}{s-s'} \|h_1\|_s \cdots \|h_p\|_s \frac{1}{r^p} \frac{1}{\eta^q},$$

$$(p > 0, \quad q \geq 0).$$

(IV) $\|F_{0,q}\|_{s'} \leq \dfrac{c}{s - s'} \dfrac{1}{\eta^q}$ $q \geq 0$ with $r, \eta, c > 0$.

There exists ([1]) only a formal series $x = \sum_{m \geq 0} x_m T^m$, where

$$x_{k+1} = \dfrac{1}{k+1} \left[F_{0,k} + \sum_{p=1}^{k} \sum_{q+n_1+\cdots+n_p=k} F_{p,q}(x_{n_1}, \ldots, x_{n_p}) \right] \quad (4)$$

when $k \geq 0$ and $x_1 = F_{0,0}$ such that

$$|t| < \Delta(1 - s) \to \sum_{n \geq 0} x_n t^n \in X_s$$

is a solution of (3) ($\forall\ 0 < s < 1$).

We have

$$\|x_n\|_s \leq \dfrac{r}{8sn^2} \left(\dfrac{1}{\Delta(1 - s)} \right)^n, \quad n \geq 1,$$

where $\Delta = \min(\eta/2,\ r/32ec)$. Then

$$x \in H_\Delta.$$

THEOREM 1 ([2]). There exists equivalence between hypotheses (I), (II), (III), (IV) and (A), (B), (C), (D) which follow.
   Let $0 < \eta_1 < \eta$, $0 < R < r$, $c_1 = c(1 - \eta_1/\eta)^{-1}(1 - R/r)$.

(A) $f(t, s)$ defined in $(|t| < \eta_1) \times V(x_0, R)$, where $V(x_0, R) = \bigcup_{0 < s \leq 1} B_s(x_0, R)$, with values in X.

(B) $f(t, x)$ G-analytic in $(|t| < \eta_1) \times B_s(x_0, R)$ with values in $X_{s'}$ ($\forall\ 0 < s' < s \leq 1$).

(C) $\|f(t, x)\|_{s'} \leq \dfrac{c_1}{s - s'} \impliedby \|x - x_0\|_s < R$,

$|t| < \eta_1$, $0 < s' < s \leq 1$.

The proofs of (A), (B), and (C) are immediate.

Conversely, let f be a function that satisfies (A), (B), and (C).

We will show (I), (II), (III), (IV), with $c = c_1$, $\eta = \eta_1$, $r = R/e$.

When $|t| < \eta$ $\|x - x_0\|_s < R$ we have

$$f(t, x) = \sum_{\substack{p \geq 0 \\ q \geq 0}} \frac{1}{p!} \delta_x^p \left[\frac{1}{q!} \delta_t^q f(0, x_0, t) \right] (x - x_0)$$

$$= \sum_{\substack{p \geq 0 \\ q \geq 0}} \frac{1}{p!} \delta_x^p \frac{1}{q!} [\delta_t^q f(0, x_0, 1)] (x - x_0) t^q.$$

Let

$$F_{p,q}(h_1, \ldots, h_p) = \frac{1}{p!} \delta_x^p \frac{1}{q!} [\delta_t^q f(0, x_0, 1)] (h_1, \ldots, h_p)$$

when $p > 0$, $q \geq 0$, and $F_{0,q} = \delta_t^q f(0, x_0, 1)(1/q!)$ ($q \geq 0$) i.e., (I) and (II). From (C) we have

$$\frac{1}{q!} \left\| \left(\frac{d^q}{dt^q} f(0, x) \right) \right\|_{s'} \leq \frac{c}{s - s'} \frac{1}{\eta^q} \Longleftarrow 0 < s' < s \leq 1,$$

$$\|x - x_0\|_s < R, \quad q \geq 0,$$

i.e., (IV) when $x = x_0$.

By our inequality ([3]), we have

$$\|F_{p,q}(h_1, \ldots, h_p)\|_{s'} \leq \frac{1}{s - s'} \frac{1}{\eta^q} \frac{p^p}{p!} \|h_1\|_s \cdots \|h_p\|_s \frac{1}{R^p}.$$

But $p^p/p! \leq e^p$ ($p \geq 1$), so we have (III) with $r = R/e$, $\eta = \eta_1$, $c = c_1$.

DEFINITION. *Analytic mapping*: a G-analytic and locally bounded mapping from an open set of a complex Banach space into another complex Banach space. *Banach local analytic manifold*: a triplet (Ω, f, Y). Ω is open in a complex Banach space X, Y is a complex Banach space, f is an analytic injective mapping from Ω in Y, with continuous inverse and differential $h \in X \to \delta_x^h f \in Y$ injective $\forall x \in \Omega$.

THEOREM 2. Solutions of a nonlinear differential equation as a local Banach manifold.

Let
$$\frac{dx}{dt} = f(t, x), \quad x(0) = \varphi. \tag{5}$$

Let f be
$$f: (|t| < \eta) \times V(0, R) \to X,$$

$$f: (|t| < \eta) \times B_s(0, r) \to X_{s'}, \quad (0 < s' < s \le 1)$$

G-analytic, and
$$\|f(t, x)\|_{s'} \le \frac{c}{s - s'} \Longleftarrow \|x\|_s < R,$$

$$|t| < \eta, \quad 0 < s' < s \le 1.$$

When $\|\varphi\|_1 < \rho < R$, we can associate, by Example 2, the solution $x(\varphi) \in H_\Delta$ with $\delta = \min(\eta/2, (R - \rho)/32e^2 c)$, for $\|x - \varphi\|_s < R - \rho \Longrightarrow \|x\|_s < R$: $x(\varphi)$ depends injectively on φ as the unique solution of the system (5) and has continuous inverse, for $\|x\| \ge \|x_0\|_1 = \|\varphi\|_1$.

We shall show that

$x(\varphi)$ is analytic and its differential in any point is an isomorphism between the Banach space X_1 and its image in H_Δ.

We have $x(\varphi) = \Sigma_{n \ge 0} x_n(\varphi) T^n$. The recurring formulas of x_n [Example 2, formula (4)] give us their G analyticity.

We have
$$\|x_n(\varphi)\|_s \le \frac{R - \rho}{8eSn^2} \left(\frac{1}{\Delta(1 - s)}\right)^n; \quad (\|\varphi\|_1 < \rho, 0 < s < 1).$$

Then

(a) $\|x(\varphi)(t)\|_s \le R$ when $0 < s < 1$, $|t| < \Delta(1 - s)$, and $\|\varphi\|_1 < R$.

(b) $\|x(\varphi)\| < R$ when $\|\varphi\|_1 < \rho$.

Then

(I) $\|\varphi\|_1 < \rho \to x(\varphi)(t) = \sum_{n \geq 0} x_n(\varphi) t^n \in X_s$, $|t| < \Delta(1 - s)$

is analytic when $0 < s < 1$.

(II) $\left\| x(\varphi + h)(t) - x(\varphi)(t) - \sum_{n \geq 0} \delta x_n(\varphi, h) t^n \right\|_s$

$$\leq R \sum_{m \geq 2} \left(\frac{\|h\|_1}{\rho - \|\varphi\|_1} \right)^m$$

when $|t| < \Delta(1 - s)$, $0 < s < 1$, $\|h\|_1 < \rho - \|\varphi\|_1$. Then

$$\left\| x(\varphi + h) - x(\varphi) - \sum_{n \geq 0} \delta x_n(\varphi, h) T^n \right\| = o(\|h\|_1),$$

i.e., the analyticity of $x(\varphi)$ [use (b)].

(III) The mapping $h \in X_1 \to \delta x(\varphi, h)$ is injective: For $\delta x(\varphi, h) = 0 \implies h = \delta x(\varphi, h)(0) = 0$. We then have the following theorem:

> The mapping from initial conditions to the solutions of system (5) defines a local analytic Banach manifold.

Example 3.

$$\frac{\partial z}{\partial t} = f\left(t, \tau_1, \ldots, \tau_n, z, \frac{\partial z}{\partial \tau_1}, \ldots, \frac{\partial z}{\partial \tau_n} \right),$$

(6)

$$z(0, \tau) = (\tau),$$

where f is a complex analytic function of variables $(t, \tau_1, \ldots, \tau_n, z, u_1, \ldots, u_n) \in \mathbb{C}^{2n+2}$ when $|t| < \eta$, $|z| < |\tau_j| < 1$,

$|u_j| < R$, bounded by $M > 0$, and φ is analytic in the ball $|\tau| < 1$, with values in $\mathbb{C}$.

Let X_s be the Banach scale of Example 1 and

$$F(t, x)(\tau) = f\left(t, \tau_1, \ldots, \tau_n, x, \frac{\partial x}{\partial \tau_1}(\tau), \ldots, \frac{\partial x}{\partial \tau_n}(\tau)\right)$$

when $|x|_s < R$, $|\tau| < s$, $0 < s \leq 1$.

We have

$$|F(t, x)(\tau)| \leq M < \frac{M}{s - s'}$$

$$\left|\frac{\partial}{\partial \tau_j} F(t, x)(\tau)\right| \leq \frac{M}{s - s'}$$

when $|x|_s < R$, $|\tau| < s' < s \leq 1$, i.e.,

$$F(t, x) \in X_{s'} \quad \text{and} \quad \|F(t, x)\|_{s'} \leq \frac{M}{s - s'}.$$

We then have the following, by Theorem 2 and Example 1:

The mapping that associates to φ, $\|\varphi\|_1 < \rho < R$, the solution $z \in H^1(B_\Delta, \mathbb{C})$ of (6) defines a local Banach analytic manifold.

REFERENCES

1. J. F. Treves, An abstract nonlinear Cauchy-Kovalewska theorem, *Trans. Amer. Math. Soc.*, 150 (1970), 77-92.
2. D. Pisanelli, Théorèmes d'Ovcyannicov, Frobenius et d'inversion et groupes de Lie locaux dans les échelles d'espaces de Banach, *C. R. Acad. Sci*, 6 (12 Nov. 1972), 277.
3. D. Pisanelli, Contribuição ao estudo dos operadores analíticos (Tese de livre-docência), *Boletim da Sociedade de Matemática de São Paulo*, 1961.
4. T. Yamanaka, Note on Kovaleskajas system of partial differential equations, *Comm. Math. Univ. Sancti Pauli, Tokyo, Rikkyo Daigoku*, 9 (1960), 7-10.

WEIGHTED APPROXIMATION AND FULL COMPLETENESS IN SPACES OF CONTINUOUS FUNCTIONS[*]

W. H. Summers[†]

Instituto de Matemática
Universidade Federal do Rio de Janeiro
Rio de Janeiro, Brazil

In comparison with its importance in the theory of normed linear spaces, the role of completeness in the general theory of locally convex spaces has been less than spectacular. When viewed in the light of Grothendieck's completeness criterion, however, the Krein-Šmulian theorem indicates a reason for this disparity and suggests an approach to strengthening the concept of completeness. The notion of a fully complete (B-complete) locally convex space is one outgrowth of this viewpoint.

Providing a setting for extending the classical Banach open mapping theorem to nonmetrizable locally convex spaces, fully complete spaces have been intensively investigated during the two decades since they were introduced by Collins [1] and

---

[*]This research was supported by National Science Foundation Grant GP-34370 and a grant to IM/UFRJ from the Financiadora de Estudos e Projetos (FNDCT/FINEP/SPPR). On leave from the University of Arkansas, Fayetteville, Arkansas, U. S. A., the author is also supported through a travel grant from the Comissão para o Intercâmbio Educacional entre os Estados Unidos da América e o Brasil (Comissão Fulbright).

[†]Present address: Department of Mathematics, University of Arkansas, Fayetteville, Arkansas.

Pták [2], [3], but one shortcoming of the development to date is the scarcity of information relating to concrete examples. It is still the case, for instance, that almost nothing is known about full completeness in even the familiar spaces of continuous functions except when implied by stronger properties; e.g., when the space is a Fréchet space. However, in a previous article [4], a small step toward remedying this state of affairs was taken by establishing a meaningful necessary condition for certain weighted spaces to be fully complete, and the purpose of the present discussion is to show that a solution of the weighted approximation problem due to Nachbin [5] can be used to extend this result to an even larger class of continuous function spaces.

1. WEIGHTED SPACES OF CONTINUOUS FUNCTIONS

The real and the complex number fields, each with its usual Euclidean topology, will be denoted by $\mathbb{R}$ and $\mathbb{C}$, respectively, while X will henceforth denote a completely regular Hausdorff topological space. Given any such space X, and taking $\mathbb{K} \in \{\mathbb{R}, \mathbb{C}\}$, we will let $C(X; \mathbb{K})$ denote the set of all continuous functions from X into $\mathbb{K}$, let $B(X; \mathbb{K})$ denote the set of all bounded functions from X into $\mathbb{K}$, and let $B_0(X; \mathbb{K})$ denote the set of all members of $B(X; \mathbb{K})$ which also vanish at infinity. For $f \in B(X; \mathbb{K})$, put $\|f\| = \sup\{|f(x)|: x \in X\}$.

We introduce a set V of nonnegative upper semicontinuous (u.s.c.) functions on X, the elements of V being referred to as *weights*. The corresponding *weighted space* $CV_0(X; \mathbb{K})$ is the locally convex (topological vector) space obtained by endowing the vector space over $\mathbb{K}$ consisting of those $f \in C(X; \mathbb{K})$ such that $fv \in B_0(X; \mathbb{K})$ for every $v \in V$ with the *weighted topology* ω_V generated by the seminorms p_v, one for each $v \in V$, defined on this space by putting $p_v(f) = \|fv\|$. Since there is clearly no loss of generality, we will hereafter assume that a set V of weights on X forms a convex cone. A base of closed absolutely convex neighborhoods (of the origin) in $CV_0(X; \mathbb{K})$ is thus formed by the sets $U_v = \{f \in CV_0(X; \mathbb{K}): \|fv\| \leq 1\}$, $v \in V$.

Many familiar spaces of continuous functions can be realized as weighted spaces, and some specific examples will be noted in Section 3. Weighted spaces were introduced by Nachbin [6] to provide the context for a more general version of the Bernstein approximation problem which he termed the weighted approximation problem [5, p. 293]. Nachbin's treatment of the real and the self-adjoint complex cases of this problem has appeared in monograph form [7], while a more recent look at the problem can be found in [8].

2. A NECESSARY CONDITION FOR FULL COMPLETENESS IN WEIGHTED SPACES

If f is a $\mathbb{K}$-valued function on the completely regular Hausdorff space X, we put $N(f) = \{x \in X: f(x) \neq 0\}$. More generally, if F is any set of $\mathbb{K}$-valued functions on X, then we put $N(F) = \bigcup_{f \in F} N(f)$. In addition, if F is a subset of X and if f is a $\mathbb{K}$-valued function on X, then $R_F(f)$ will denote the restriction of f to F.

In the following development, V will continue to denote a convex cone of weights on X. Moreover, if F is a subset of X, then we will let $V^F = R_F(V)$, while the restriction of R_F to $CV_0(X; \mathbb{K})$ will also be denoted by R_F. The following result is an immediate consequence of the definitions.

<u>2.1 LEMMA</u> [4, p. 1,199]. If F is a closed subset of X, then V^F is a convex cone of weights on F and R_F is a continuous linear mapping from $CV_0(X; \mathbb{K})$ into $CV_0^F(F; \mathbb{K})$.

2.2 LEMMA. If F is a closed subset of X and if $N(V) \subseteq N(CV_0(X; \mathbb{K}))$, then $R_F(CV_0(X; \mathbb{K}))$ is dense in $CV_0^F(F; \mathbb{K})$.

Proof. In view of Lemma 2.1, $W = \{R_F(f): f \in CV_0(X; \mathbb{K})\}$ is a vector subspace of the weighted space $CV_0^F(F; \mathbb{K})$. Since $CV_0(X; \mathbb{K})$ is a module over $C_b(X; \mathbb{K}) = C(X; \mathbb{K}) \cap B(X; \mathbb{K})$, it is obvious that W is a module over the subalgebra $A = \{R_F(f): f \in C_b(X; \mathbb{K})\}$ of $C(F; \mathbb{K})$; i.e., $AW \subseteq W$, where all algebraic operations are, of course, defined pointwise. Moreover, A is self-adjoint in the complex case ($\mathbb{K} = \mathbb{C}$), and each $a \in A$ is bounded on F. Taking $f \in CV_0^F(F; \mathbb{K})$, we fix $x \in F$ and $v \in V$. If $v(x) \neq 0$, then there exists a $g \in CV_0(X; \mathbb{K})$ such that $g(x) \neq 0$ whereby there is a $w \in W$ for which $w(x) = f(x)$. Hence, in any event, there exists a $w \in W$ such that $|f(x) - w(x)| v(x) = 0$. Combining this with the fact that A separates the points of F, Nachbin's solution of the weighted approximation problem in the bounded case [5, p. 295] implies that f belongs to the closure of W in $CV_0^F(F; \mathbb{K})$, and the proof is complete.

Before stating the next lemma, a definition is needed: a linear mapping t from a topological vector space E into a topological vector space F is called *nearly open* provided the closure of $t(U)$ is a neighborhood in F whenever U is a neighborhood in E.

2.3 LEMMA. If F is a closed subset of X and if $N(V) \subseteq N(CV_0(X; \mathbb{K}))$, then R_F is a continuous nearly open linear mapping from $CV_0(X; \mathbb{K})$ into $CV_0^F(F; \mathbb{K})$.

Proof. By Lemma 2.1, R_F is a continuous linear mapping from $CV_0(X; \mathbb{K})$ into $CV_0^F(F; \mathbb{K})$, while $W = \{R_F(f): f \in CV_0(X; \mathbb{K})\}$ is dense in $CV_0^F(F; \mathbb{K})$ by Lemma 2.2. Fixing $v \in V$, let $u = R_F(v)$, and put $U = \{f \in CV_0^F(F; \mathbb{K}): \|fu\| < 1\}$. Since U is an open neighborhood in $CV_0^F(F; \mathbb{K})$, $U \subseteq \overline{U \cap W}$. Further, if $f \in CV_0(X; \mathbb{K})$ with

$R_F(f) \in U$, then a straightforward argument verifies that there exists a $g \in U_v$ for which $R_F(g) = R_F(f)$; i.e., $U \cap W \subseteq R_F(U_v)$. Since U is thus contained in the closure of $R_F(U_v)$, R_F is nearly open.

Let E be a locally convex Hausdorff space. A subset A of E', the space of all continuous linear functionals on E, is called *nearly closed* if, for each neighborhood U in E, $U^0 \cap A$ is $\sigma(E', E)$-closed. If every nearly closed vector subspace of E' is $\sigma(E', E)$-closed, then E is said to be *fully complete* or *B-complete*. Fully complete spaces and related concepts have been treated in a monograph by Husain [9]. In addition, Dostal [10] and Robertson [11], among others, have discussed more recent results, techniques, and still open questions, as well as having provided extensive bibliographies.

2.4 THEOREM. If $CV_0(X; \mathbb{K})$ is fully complete and if $V \subseteq B_0(X; \mathbb{K})$, then X is normal.

Proof. It will suffice to show that if F is a closed subset of X, then F is C^*-embedded in X [12, p. 48]; i.e., if $f \in C_b(F; \mathbb{R})$, then there exists a $g \in C_b(X; \mathbb{R})$ such that $R_F(g) = f$. To this end, we fix a closed subset F of X, and let $f \in C_b(F; \mathbb{R})$. Since $C_b(X; \mathbb{K}) \subseteq CV_0(X; \mathbb{K})$, we can apply Lemma 2.3 to conclude that R_F is a continuous nearly open linear mapping from $CV_0(X; \mathbb{K})$ into $CV_0^F(F; \mathbb{K})$. Pták's open mapping theorem (cf. [4, p. 1,197]) thus implies that R_F is open onto $CV_0^F(F; \mathbb{K})$, and hence, noting that $f \in CV_0^F(F; \mathbb{K})$, there is a $g \in CV_0(X; \mathbb{K})$ for which $R_F(g) = f$. This concludes the proof, since it can clearly be assumed that $g \in C_b(X; \mathbb{R})$.

The preceding theorem considerably strengthens an earlier result [4, p. 1,200] in which rather severe restrictions were imposed on either the supports of elements in V or on the space

X itself. We take this opportunity to thank J. B. Prolla for his helpful comments relating to the application of Nachbin's approximation theorem for the bounded case of the weighted approximation problem.

3. APPLICATIONS AND OBSERVATIONS

If X is locally compact, then $C_0(X; \mathbb{K}) = C(X; \mathbb{K}) \cap B_0(X; \mathbb{K})$ with the supremum norm topology σ is a Banach space and hence is fully complete by the Krein-Šmulian theorem. Since $(C_0(X; \mathbb{K}), \sigma) = CV_0(X; \mathbb{K})$ whenever V is taken to be the convex cone of nonnegative constant functions on X and since X can be locally compact without being normal, assuming in Theorem 2.4 that $V \subseteq B_0(X; \mathbb{K})$ is, in some sense, necessary. Obviously, $V \subseteq B_0(X; \mathbb{K})$ if, and only if, $C_b(F; \mathbb{R}) \subseteq CV_0^F(F; \mathbb{K})$ for every closed subset F of X.

Collins [1, p. 275] has shown that the image of a fully complete space by a continuous open linear mapping is likewise fully complete. Therefore, the proof of Theorem 2.4 also yields the following result.

3.1 THEOREM. If $CV_0(X; \mathbb{K})$ is fully complete and if $N(V) \subseteq N(CV_0(X; \mathbb{K}))$, then, for each closed subset F of X, $CV_0^F(F; \mathbb{K})$ is fully complete.

The convex conic hull of the set of characteristic functions of all compact subsets of X is a set of weights on X, and $CV_0(X; \mathbb{K})$ in this case is $C(X; \mathbb{K})$ with the compact-open topology κ. The next theorem, implicit in an article by Pták [2, p. 345], is an immediate consequence of Theorems 2.4 and 3.1.

3.2. THEOREM. If $(C(X; \mathbb{K}), \kappa)$ is fully complete, then X is normal and $(C(F; \mathbb{K}), \kappa)$ is fully complete for every closed subset F of X.

If V is chosen to be the set of all nonnegative u.s.c. functions in $B_0(X; \mathbb{K})$, then V is a convex cone of weights on X and $CV_0(X; \mathbb{K})$ coincides with the space $C_b(X; \mathbb{K})$ equipped with the substrict topology β_0 (see [13, p. 316]). Although our earlier versions of Theorems 2.4 and 3.1 were not applicable to this important case, a result analogous to the preceding one is only known example where $(C_b(X; \mathbb{K}), \beta_0)$ is fully complete occurs when X is discrete [14, p. 367]. It would be of considerable interest to discover a necessary and sufficient condition on X for

3.3 THEOREM. If $(C_b(X; \mathbb{K}), \beta_0)$ is fully complete, then X is normal and $(C_b(F; \mathbb{K}), \beta_0)$ is fully complete for every closed subset F of X.

If X is compact, then $(C_b(X; \mathbb{K}), \beta_0)$ is a Banach space [13] and therefore fully complete. However, if X is not compact, then β_0 is not metrizable [13, p. 321], and in this case the only known example were $(C_b(X; \mathbb{K}), \beta_0)$ is fully complete is when X is discrete [14, p. 367]. It would be of considerable interest to discover a necessary and sufficient condition on X for $(C_b(X; \mathbb{K}), \beta_0)$ (or $(C(X; \mathbb{K}), \kappa)$) to be fully complete. Theorem 2.4 is one step in this direction, as is our next and concluding observation drawn from Theorem 3.3 (Theorem 3.2).

3.4 COROLLARY. If a topological property of X is necessary for $(C_b(X; \mathbb{K}), \beta_0)$ (respectively, $(C(X, \mathbb{K}), \kappa)$) to be fully complete, then this property is hereditary with respect to closed subsets of X.

REFERENCES

1. H. S. Collins, Completeness and compactness in linear topological spaces, *Trans. Amer. Math. Soc.*, 79 (1955), 256-280.
2. V. Pták, On complete topological linear spaces, *Czechoslovak Math. J.*, 78 (1953), 301-364.
3. V. Pták, Completeness and the open mapping theorem, *Bull. Soc. Math. France*, 86 (1958), 41-74.
4. W. H. Summers, Full-completeness in weighted spaces, *Canad. J. Math.*, 22 (1970), 1,196-1,207.
5. L. Nachbin, Weighted approximation for algebras and modules of continuous functions: real and self-adjoint complex cases, *Ann. Math.*, 81 (1965), 289-302.
6. L. Nachbin, Weighted approximation over topological spaces and the Bernstein problem over finite dimensional vector spaces, *Topology*, 3 (1964), Suppl. 1, 125-130.
7. L. Nachbin, *Elements of Approximation Theory*, Math. Studies 14, Van Nostrand, Princeton, 1967.
8. W. H. Summers, Weighted approximation for modules of continuous functions II, *Analyse Fonctionnelle*, C. R. Colloque d'Analyse (Rio de Janeiro, 1972), Hermann, Paris, 1975.
9. T. Husain, *The Open Mapping and Closed Graph Theorems in Topological Vector Spaces*, Oxford Univ. Press, Oxford, 1965.
10. M. A. Dostal, Some recent results on topological vector spaces, *Functional Analysis and Applications*, Proc. Symp. Analysis (Recife, 1972); *Lecture Notes in Mathematics*, 384, Springer-Verlag, New York, 1974, 20-91.
11. W. Robertson, On the closed graph theorem and spaces with webs, *Proc. London Math. Soc.*, 24 (1972), 692-738.
12. L. Gillman and M. Jerison, *Rings of Continuous Functions*, Van Nostrand, Princeton, 1960.
13. F. D. Sentilles, Bounded continuous functions on a completely regular space, *Trans. Amer. Math. Soc.*, 168 (1972), 311-336.
14. H. S. Collins, On the space $\ell^\infty(S)$, with the strict topology, *Math. Z.*, 106 (1968), 361-373.

AN INVITATION TO THE THEORY OF ORDINARY DIFFERENTIAL EQUATIONS IN BANACH SPACES

Giovanni Vidossich[†]

Departamento de Matemática
Universidade de Brasília
Brasília, Brazil

The purpose of this discussion is to give an informal view of the theory of ordinary differential equations in Banach spaces in hopes of arousing an interest in this area among young mathematicians. In my opinion the present stage of the theory may be classified as "a good start" and nothing more. Therefore this theory needs new energies in order to solve the various problems and to progress to a rich stage. This field is interesting because it is difficult, and is important because it has many applications. Perhaps the best way to accept these claims is to look at the four fundamental occurrences during of its history. For this reason I will speak about each one of them. However, the discussion will be informal, since the technical details are so involved they may detract from the overall picture.

[†]Present address: Istituto di Matematica, Università di Trieste, Trieste, Italy.

1. THE FIRST HAPPENING: DIEUDONNÉ'S EXAMPLE

I believe it was Fréchet who had first the idea of doing ordinary differential equations in Banach spaces. The first important publication was a book published in France in the 1930s which dealt mainly with the linear case. The second publication of some importance was the first edition of Bourbaki [1] which considered the nonlinear case, although these nonlinearities dealt with the case of locally Lipschitz functions. This is an important but simple case. Therefore, at this point a natural question is: "Why in many years had mathematicians reached only this elementary stage of the theory?" This question was answered by Dieudonné in 1950. In his paper [2], Dieudonné shows by an example that the theorem of Peano cannot be extended to differential equations in Banach spaces of infinite dimension. His example is very interesting, since a link is made with ordinary equations in the real line and the theory of comparison for solutions of ordinary differential equations in R is employed. Dieudonné considered the Banach space

$$c_0 = \{x = (x_n)_n \mid x_n \in \mathbb{R}, \lim_n x_n = 0\}$$

with the sup norm

$$\|x\| = \sup_n |x_n|.$$

Define $f: c_0 \to c_0$ by

$$f(x) = \left(|x_n|^{1/2} + \frac{1}{n}\right)_{n=1}^{\infty}$$

and let us consider the Cauchy problem

$$x' = f(x), \quad x(0) = 0. \tag{1}$$

It is easily seen that f is continuous (and uniformly continuous on bounded sets). If (1) has a solution x defined in some interval $[0, \delta[$, then from

$$x'(t) = f(x(t))$$

it follows, putting $x(t) = (x_n(t))_n$,

$$x_n' = |x_n|^{1/n} + \frac{1}{n}, \quad x_n(0) = 0, \tag{2}$$

using the fact that $x \mapsto x_n$ (the n-th canonical projection $c_0 \to \mathbb{R}$) is continuous. In other words, if a solution of (1) exists, then its coordinate functions must satisfy an equation in the real line, namely (2). Since

$$|u|^{1/n} + \frac{1}{n} > |u|,$$

a well-known comparison theorem shows that x_n is greater than or equal to all solutions of the ordinary equations

$$u' = |u|^{1/2}, \quad u(0) = 0. \tag{3}$$

Since $u(t) = t^2/4$ solves (3), we have

$$\frac{t^2}{4} \leq x_n(t) \quad (t \in [0, \delta[; \; n \in Z^+).$$

Then for $t_0 \in [0, \delta[$, $\lim_n x_n(t_0) > 0$ so that $(x_n(t_0))_n \notin c_0$, which is a contradiction. We conclude that (1) cannot have a solution.

This example marked the first fundamental development of the theory. It established that the problem is very difficult, and therefore various mathematicians were interested in obtaining a satisfactory solution of the Cauchy problem. After Dieudonné's paper up to this date I think there have appeared some 30 papers with a sufficient condition for the existence of a solution of Cauchy problem. However, none of these conditions is totally satisfactory, although various are interesting because of some particular feature. All these theorems can be divided in two classes: those which use some compactness assumption and those which generalize the Lipschitz case. While the later use a differential inequality technique, the former usually employ a fixed point technique (of condensing mappings, in the most recent papers). Among the former, perhaps the nicest statements were given by Ambrosetti [3] while the most general seems to be Cellina [4]. The second class of theorems were inspired by the theory of monotone operators. The founder was a theorem of Browder [5], which was generalized by several authors. One of the methods used (cf. Vidossich [6]) is to approximate the given function f by locally Lipschitz maps f_n and then to show that the sequence $(x_n)_n$ of solutions of

$$x_n' = f(t, x_n), \quad x_n(t_0) = x_0$$

is a Cauchy sequence. Therefore the sequence converges to a solution of the differential equation. In proving that $(x_n)_n$ is a Cauchy sequence, differential inequalities are used.

In closing this section, let us remark that for some time there was the conjecture that continuity, or at least uniform continuity, would imply existence of solutions for the Cauchy problem in Hilbert spaces. This is not true, as was shown by Yorke [7] in 1970.

2. THE SECOND OCCURRENCE: DISCOVERY OF APPLICATIONS TO PARTIAL DIFFERENTIAL EQUATIONS

The second important development in the theory occurred around 1960. It was the discovery that ordinary differential equations in suitable Banach spaces provide an excellent setting for solving problems linked with partial differential equations. For linear equations, the fact was discovered by various people. One of the most representative among the pioneering works is Agmon and Niremberg [8]. For more recent works, compare the schools from France, Israel, Japan, the United States, and Russia. At the beginning of the 1960s, nonlinear equations were also solved by this method. The pioneering works for the nonlinear case are Prodi [9] and Browder [10], as far as I know. Problems such as uniqueness, existence, regularity, and asymptotic behavior of solutions can be treated very well by using ordinary differential equations in Banach spaces. A peculiar feature of the ordinary differential equations associated with partial differential equations is that the second member is not defined in an open set; it is defined on a dense vector subspace. In general, the equations have the form

$$x' = A(t)x + f(t, x)$$

where $A(t)$ is a linear operator defined in a dense vector subspace, and $f: \mathbb{R}^+ \times U \to X$ with $U \subsetneq X$.

I will now illustrate this topic by two examples, on uniqueness and on existence problems respectively. The subject is very difficult, and I will only show the general underlying ideas, avoiding the technical details as well as the most general setting of the results.

The first example is a uniqueness result for parabolic equations. Consider the equation

$$u_t = u_{xx} + f(t, x, u) \qquad (4)$$

with f a real-valued continuous functions defined on the set

$$\{(t, x) \in \mathbb{R}^2 \mid a \leq t \leq b, \alpha(t) \leq x \leq \beta(t)\} \times \mathbb{R},$$

where $\alpha, \beta: [a, b] \to \mathbb{R}$ are C^1-functions such that $\alpha < \beta$. I want to show that (4) has at most one solution u such that the support of $u(t, \cdot)$ is contained in $[\alpha(t), \beta(t)]$ and $u(0, x) = g(x)$, where g is a given function in $L^2(\mathbb{R})$, when there exists a constant K such that

$$f(t, x, u) - f(t, x, v) \leq K(u - v) \qquad (u < v).$$

Therefore the result applies in the particular cases when f is Lipschitz or $f(t, x, \cdot)$ is increasing. Let X be the Banach space $L^2(\mathbb{R})$. For every $t \in [a, b]$, define

$$D(t) = \{v \in C^2(\mathbb{R}) \mid \text{supp}(v) \subseteq [\alpha(t), \beta(t)]\}.$$

Let $A \subseteq \mathbb{R} \times X$ be the set

$$A = \bigcup_{a \leq t \leq b} \{t\} \times D(t).$$

Define $F: A \to X$ by

$$F(t, v) = v''(\cdot) + f(t, \cdot, v(\cdot)).$$

It can be proved that for every solution u of (4) which satisfies the given boundary condition, the function $\bar{u}: [a, b] \to X$ defined by

$$\bar{u}(t) = u(t, \cdot)$$

is a solution of the Cauchy problem

$$\bar{u}' = F(t, \bar{u}), \quad \bar{u}(0) = g \qquad (5)$$

in the Banach space X (for details, see Ladas and Lakschmikantham [11, p. 138]). Now define $V: [a, b] \times A \times A \to \mathbb{R}$ by

$$V(t, v, w) = e^{-2Kt} \int_{\alpha(t)}^{\beta(t)} |v(x) - w(x)|^2 \, dx.$$

It is easily seen that V has the following properties:

(i) $V(t, v, w) > 0$ if $v \neq w$, and $V(t, v, v) = 0$;

(ii) $\frac{\partial}{\partial t} V(t, v, w) + \frac{\partial}{\partial v} V(t, v, w) F(t, v)$

$\qquad + \frac{\partial}{\partial w} V(t, v, w) F(t, w) \leq 0.$

Therefore we have the uniqueness of (5), hence of (4), simply by applying a theorem of Murakami [12].

The second example is an existence theorem for wave equations. The Cauchy problem

$$u_{tt} = \Delta u - ku^3, \quad u(0, x) = \varphi(x), \quad \frac{\partial}{\partial t} u(0, x) = \psi(x), \qquad (6)$$

where $\varphi \in W^{2,2}$ and $\psi \in W^{1,2}$ are given, was studied by Jörgens [13] with analytical techniques. As this problem is of interest in quantum theory, Browder got into the problem and succeeded in associating with (6) a second-order Cauchy problem

$$v'' = v_{xx} - kv^3, \quad v(0) = \varphi, \quad v'(0) = \psi \qquad (7)$$

in a suitable Banach space. The solutions of (6) and (7) are related by the relation

$$v(t)(x) = u(t, x).$$

His argument is very involved, and I will give only an idea. The interested reader may look at the original source, Browder [10]. Let $H = L^2(\mathbb{R}^n)$ and let X be the set of all $v \in C^2([0,T], H)$ such that $v(t) \in W^{m,2}$ for all t, $v \in C^1([0, T], W^{m,2})$, and Δv is continuous. Browder shows that X is complete by the norm

$$\|v\|_X = \sup_{0 \le t \le T} \{\|v''(t)\|_H + \|\Delta v(t)\|_H + \|v'(t)\|_{W^{m,2}} + \|v(t)\|_{W^{m,2}}\}.$$

Moreover, Browder shows that for every $f \in X$ there is a unique $v_f \in X$ such that

$$v_f'' = \Delta v_f + f(t), \quad v(0) = \varphi, \quad v'(0) = \psi.$$

The function v_f satisfies

$$\|v_f\|_X \le c\{\|\Delta\varphi\|_H + \|\varphi\|_H + \|\varphi\|_{W^{m,2}} + \|\psi\|_{W^{m,2}}$$

$$+ T \sup_{0 \le t \le T} (\|f(t)\| + \|f'(t)\|)\} \qquad (8)$$

where c is a constant independent of φ, ψ, f, and T. Using this estimate, Browder shows the existence of $\varepsilon > 0$ such that the operator U: f $\mapsto$ v$_f$ maps into itself the ball B in X of center 0 and radius ε, and U is a contraction: B $\to$ B. The first claim follows from the fact that the last two addenda in (8) are small with T, while the second claim follows from (8) applied to f - g (hence with $\varphi = \psi = 0$) and the fact that t $\mapsto$ -kt^3 is Lipschitz on bounded sets. By the Banach-Caccioppoli fixed point theorem, U has a fixed point. This point is a solution of (7), hence of (6), in [0, T]. The argument of Browder works also for more general cases, as can be seen in [10].

Nonlinear partial differential equations which have been studied successfully so far by using ordinary differential equations in Banach spaces are those of parabolic type and those of Navier-Stokes (compare the results of Prodi's and the Israel schools). Finally, we remark that related to this subject there appeared in the last two years two remarkable papers, namely, Crandall [14] and Bourguignon and Brézis [15]. The particular feature of these papers is that both use some of the latest results of the theory of ordinary differential equations in Banach spaces. Crandall treats first-order partial differential equations by using nonlinear semigroups, while Bourguignon and Brézis deal with the Euler equation on the basis of an existence theorem due to Martin [16] for Cauchy problems associated with ordinary differential equations on closed sets of Banach spaces.

3. THE THIRD OCCURRENCE: THE DISCOVERY OF APPLICABILITY TO NONLINEAR FUNCTIONAL ANALYSIS

The third important happening was the discovery, around 1965, that the theory of ordinary differential equations in Banach spaces can be used conveniently in Nonlinear Functional

Analysis to obtain fixed point theorems or zero theorems, i.e., to solve the equation

$$F(x) = 0. \qquad (9)$$

Perhaps Micheletti in [17] was the first who proposed an idea to solve equation (9) by using ordinary differential equations. She proved a surjectivity theorem for monotone operators using the asymptotic behavior for solutions of an ordinary differential equation. To date we know four different methods to solve (9) by using the theory of ordinary differential equations in Banach spaces:

(a) The method based on asymptotic behavior of solutions.

(b) The method based on uniqueness of solutions of Cauchy problems.

(c) The method based on the existence of periodic solutions with small periods.

(d) The continuation method.

Methods (a), (b), and (c) work well above all for monotone and accretive operators. Since they are treated in Vidossich [18], I will avoid speaking of them here. I remark only that the theorems proved by these methods have not been proved by a different method, as far as I know. I will speak only about (d). This method applies to operator F which are of class C^1. Although this is a restriction on the generality of the results, the C^1 operators have applications in various problems, as for example those arising in the theory of elliptic equations.

Let X, Y be Banach spaces, $U \subseteq X$ open, and $F: U \to Y$ of class C^1. The argument I am going to explain can be looked at as inspired by the homotopy invariance of the topological degree. Fix $x_0 \in U$ and consider the homotopy $H: [0, 1] \times U \to Y$ defined by

$$H(t, x) = F(x) - (1 - t) F(x_0) \qquad (10)$$

Suppose that for every $t \in [0, 1]$ there exists $x_t \in U$ such that $H(t, x_t) = 0$. Then $x: t \mapsto x_t$ is a function $[0, 1] \to U$. Suppose x is differentiable. By differentiating both members of

$$0 = F(x(t)) - (1 - t) F(x_0)$$

we obtain

$$F'(x(t)) x'(t) = -F(x_0). \qquad (11)$$

If for every t the linear operator $F'(x(t))$ has a right-inverse operator $F^*(x(t))$, then (11) implies

$$x'(t) = -F^*(x(t)) F(x_0).$$

Now let us study the Cauchy problem

$$x' = -F^*(x) F(x_0), \quad x(0) = x_0. \qquad (12)$$

Suppose that x is a solution of (12) on $[0, 1]$. Applying $F(x(t))$ to both members of equation (12) we get (11). Taking $\int_0^t$ of both members of (11) we get (10). Taking $t = 1$ in (10) we get $F(x(1)) = 0$. Therefore, the entire problem is reduced to finding global solutions of (12). A typical theorem obtained by this method is the following.

THEOREM (Fabbry [19]): Let X, Y be Banach spaces, $B = B(x_0, \varepsilon)$ a ball of X, and $F: \bar{B} \to Y$ Fréchet differentiable in B. Suppose each $F'(x)$ has a right inverse $F^*(x)$ and that a continuous function $\omega: [0, \varepsilon] \to \mathbb{R}^+$ exists such that

$$\|F^*(x) F(x_0)\| \leq \omega(\|x - x_0\|).$$

If $x \mapsto F^*(x) F(x_0)$ is locally Lipschitz in B and if the equation

$$\int_0^r \frac{du}{\omega(u)} = 1$$

has a solution $r^* \in [0, \varepsilon]$, then $F(x) = 0$ has a solution.

I close this section by noting that the basic ideas of this method were considered earlier by Wazowski [20] to prove an implicit function theorem.

4. THE FOURTH OCCURRENCE: THE DISCOVERY OF APPLICABILITY TO FUNCTIONAL EQUATIONS

The last fundamental happening is very recent, occuring after 1970. Lasota and Yorke observed first in [21] that some problems of functional equations, and in particular of delay equations, can be solved by using an ordinary differential equation in a suitable Banach space. Their idea has been used by Li [22], while recently Nussbaum [23] solved in this way a problem which was left open by Jones years ago. I will show the basic ideas of this argument by illustrating the proof of the following theorem, which includes as special cases two results of Lasota and Yorke [21]:

THEOREM (Li [22]). If x is a nonconstant, p-periodic solution of $x' = f(x_t)$ and if there exists $g: [-h, 0] \to \mathbb{R}^+$ such that

$$|f(u) - f(v)| \leq \int_{-h}^{0} g(\xi)|u(\xi) - v(\xi)|\, d\eta(\xi)$$

where η is increasing and bounded, then

$$p \geq \frac{2\pi}{\int_{-h}^{0} g(\xi)\, d\eta(\xi)}.$$

In the equation $x' = f(x_t)$ it is understood that $f: C([-h, 0]) \to \mathbb{R}$ and is equivalent to $x' = F(x(t - h))$. To prove the above theorem, consider the Hilbert space $H = L^2_{\#}(\mathbb{R})$ of p-periodic functions in $L^2(\mathbb{R})$ with the scalar product

$$(u|v) = \int_0^p u(\xi) \, v(\xi) \, d\xi.$$

Define $G: H \to H$ by

$$G(x)(s) = f(x_s),$$

where

$$x_s(t) = x(s + t) \quad (-h \leq t \leq 0).$$

Let x be a p-periodic solution of $x' = f(x_t)$. It can easily be proved that the function $y: \mathbb{R} \to H$ defined by

$$y(t)(x) = x(t + s)$$

is Fréchet differentiable and that its derivative satisfies $y'(t)(s) = x'(t + s)$. From this we derive that $y' = G(y)$. Therefore, in order to get the conclusion, we could apply the following result—

THEOREM (Lasota and Yorke [21]). Let H be a Hilbert space, $f: A \subseteq H \to H$ Lipschitz with constant L. If there exists a nonconstant, p-periodic solution of $x' = f(x)$, then $p \geq 2\pi/L$.

—if we could prove that G is Lipschitz of constant $\leq \int_{-h}^{0} g(\xi) \, d\eta(\xi)$. This is done in the paper of Li.

REFERENCES

1. N. Bourbaki, *Fonctions d'une Variable Réelle* (1st ed.), Hermann, Paris, 19 , Chapter 4.
2. J. Dieudonné, Deux examples singuliers d'équations différentielles, *Acta Sci. Math. (Szeged)*, 12 (1950), 38-40.
3. A. Ambrosetti, Un teorema di esistenza per le equazioni differenziali negli spazi di Banach, *Rend. Sem. Mat. Padova*, 39 (1967), 349-360.
4. A. Cellina, On the existence of solutions of ordinary differential equations in Banach spaces, *Funkcial. Ekvac.*, 14 (1971), 129-136.
5. F. E. Browder, Nonlinear equations of evolution, *Ann. Math.* 80 (1964), 485-523.
6. G. Vidossich, Existence, comparison and asymptotic behavior of solutions of ordinary differential equations in finite and infinite dimensional Banach spaces, *Notas de Matemática*, No. 24, Universidade de Brasília, 1972.
7. J. A. Yorke, A continuous differential equation in Hilbert space without existence, *Funkcial. Ekvac.*, 13 (1970), 19-21.
8. S. Agmon and L. Nirenberg, Properties of solutions of ordinary differential equations in Banach space, *Comm. Pure Appl. Math.*, 16 (1963), 121-239.
9. G. Prodi, Teoremi di tipo locale per il sistema di Navier-Stokes e stabilità delle soluzioni stazionarie, *Rend. Sem. Math. Univ. Padova*, 32 (1962), 374-397.
10. F. E. Browder, On nonlinear wave equations; *Math. Z.*, 80 (1962), 249-264.
11. G. E. Ladas and V. Lakshmikantham, *Differential Equations in Abstract Spaces*, Academic Press, New York, 1972.
12. H. Murakami, On non-linear ordinary and evolution equations, *Funkcial. Ekvac.*, 9 (1966), 151-162.
13. K. Jörgens, Das Anfangswertproblem in Grossen für eine Klasse nichtlinearer Wellengleichungen, *Math. Z.*, 77 (1961), 295-308.
14. M. G. Crandall, The semigroup approach to first-order quasilinear equations in several space variables, *Israel J. Math.*,
15. J. P. Bourguignon and H. Brézis, Remark on the Euler equation, *J. Funct. Anal.*, 15 (1974), 341-363.
16. R. H. Martin, Differential equations on closed subsets of a Banach space, *Trans. Amer. Math. Soc.*, 179 (1973), 399-414.

17. A. M. Micheletti, Un'osservazione sulla teoria degli operatori monotoni, *Boll. Un. Mat. Ital.*, *1* (1968), 554-558.

18. G. Vidossich, How to get zeros of nonlinear operators using the theory of ordinary differential equations, *Atas Semana Anal. Func. Não Linear, São Paulo, 1973*, edited by Soc. Bras. Mat. and by I.M.P.A., Rio de Janeiro, 1974.

19. C. Fabry, The continuation method in Banach and Hilbert spaces, in *Equations différentielles et* ... (Janssens, Mawhin, and Rouche, eds.), Hermann, Paris, 1973, pp. 345-356.

20. T. Wazewski, Sur l'evaluation du domain d'existence des fonctions implicites réelles ou complexes, *Ann. Polon. Math.*, *20* (1947), 81-125.

21. A. Lasota and J. A. Yorke, Bounds for periodic solutions of differential equations in Banach spaces, *J. Diff. Eq.*, *10* (1971), 83-91.

22. T. Y. Li, Bounds for the periods of periodic solutions of differential delay equations, *J. Math. Anal. Appl.* (to appear).

23. R. Nussbaum, Periodic solutions of analytic functional differential equations are analytic, *Michigan Math. J.*, *20* (1973), 249-255.

INDEX

A

Atomic, 197

B

Banach scale, 291
Bicharacteristic strip, 40
Bilinear associative system (BAS), 144
Bilinear triple (BT), 138
B Variation, 142
Bounded variation, 197

C

Canonical transformations, 42
Caratheodory condition, 96
Catastrophe, 111
    butterfly, cusp, fold, swallowtail, 126
    elementary, 113
    set, 126
    umbilic, 129
C^∞ local solvability, 80
Commutative algebra, 114
Condition (FZ), 40
Condition (Ψ), 40
Continuous component, 163
Control Variables, 112
Currents, 74
    Integral, 76
    normal, 77
    quasi-normal, 77
    Rectifiable, 75

D

Darboux condition, 160
Differential inequalities, 312
Differential topology, 114
Dirichlet formula, 135
Discontinuous change, 112
D' local solvability, 80

E

Equation
    functional, 319
    functional differential, 195
    hyperbolic, 191
    linear autonomous, 202
    neutral type, 191
    ordinary differential, 309
        in Banach spaces, 309
        in Hilbert spaces, 312
    parabolic, 313
    retarded differential, 191
    wave, 315

F

Fourier integral operator, 44

G

Germs (of C^∞-functions, 144
    codimension, 116
    degenerate, 114
    right-determined, 116
    singluar, 114
Girth, 216
Girth curve, 218
Group
    analytic transformation, 238
    automorphism, 235
    Lie, 238
    strongly continuous transformation, 239

H

Hausdorff, 72
    area, 72

(Hausdorff)
    measure, 72
Holomorphic Banach-Steinhaus
    theorem, 266
Hypoellipticity, 88

I

Infectious, 192
Infectious disease, 192
Infinite supported tree
    property, 221
Infinite tree property, 221
Integral
    geometric measure, 72
    interior, 139
    interior upper, 162
    Stieljes, 202
    upper, 162

J

Josefson-Nissenzveig theorem, 274
Jump component, 163

K

k-jet, 116
Kolmogorov n-width, 230

L

Local analytic manyfold, 297
Local solvability, 80
Locally solvable, 41
Lyapunov functionals, 209
Lyapunov Theory, 209

M

Mapping
    analytic, 297
    of monotone type (with respect to two Banach space), 92
    of type (M), 93
    quasi-bounded, 93
Maximum-minimum characterization of eigenvalues, 226
Measles, 192

Minimum-maximum characterization of eigenvalues, 226
Mizohata-Ohya invariant, 41
Montel Property, 246

N

Neumann problem, 92
Non atomic, 197

O

Oscillation, 137
    interior, 137

P

Principal symbol, 40
Principal type, 40
Pseudodifferential operator, 39

R

Razumikhin theorem, 213
Rectifiable sets, 71
Regulated function, 142
Riesz representation theorem, 196
Right-equivalence, 115

S

Schauder basis, 224
Semi-coercive problem, 92
Semivariation, 143
Singularity, 111
Space
    completely flat Banach, 217
    flat Banach, 216
    Frechet-Schwartz, 81
    holomorphically barreled, 265
    holomorphically barreled, locally convex, 247
    holomorphically bornological, 256
    holomorphically infrabarreled, 268
    holomorphically infrabarreled locally convex, 247

INDEX

(Space)
    holomorphically Mackey, 272
    ordered topological, 9
    Silva, 83
    Weak Zorn, 248
    Zorn, 248
Splitting Lemma, 119
Stable
    globally uniform asymptotically, 208
    uniformly, 209
    uniformly asymptotically, 208
Stone-Weierstrass Theorems, 9
Strongly nonlinear operator, 92
Subelliptic, 41
Substitution formula, 135

T

Tietze extension theorem, 200

U

Unfolding, 111
    codimension, 120
    equivalence, 121
    induced, 121
    universal, 113
    versal, universal, k-transversal, 122

V

Variation, 143

W

Weak variation, 143

Z

Zeros of nonlinear operators, 317

QA
320
S93
1974

FEB 18 1977